PLUMBER'S LICENSING STUDY GUIDE

About the Author

R. Dodge Woodson is a Master Plumber and Master Gasfitter with over 30 years in the plumbing business. He has owned and operated plumbing and construction companies in Virginia and Maine, and has served as adjunct faculty at the Central Maine Technical College, where he taught code and apprenticeship classes. Mr. Woodson is the author of numerous books, including the *2006 International Plumbing Codes Handbook, National Plumbing Codes Handbook, and International* and *Uniform Plumbing Codes Handbook,* all from McGraw-Hill.

PLUMBER'S LICENSING STUDY GUIDE

SECOND EDITION

R. Dodge Woodson

McGraw-Hill

New York Chicago San Francisco Lisbon London Madrid
Mexico City Milan New Delhi San Juan Seoul
Singapore Sydney Toronto

Copyright © 2007, 2002 by The McGraw-Hill Companies, Inc. All rights reserved. Printed in the United States of America. Except as permitted under the United States Copyright Act of 1976, no part of this publication may be reproduced or distributed in any form or by any means, or stored in a data base or retrieval system, without the prior written permission of the publisher.

1 2 3 4 5 6 7 8 9 0 QPD/QPD 0 1 2 1 0 9 8 7 6

ISBN-13: 978-0-07-147939-4
ISBN-10: 0-07-147939-2

The sponsoring editor for this book was Larry S. Hager and the production supervisor was Pamela A. Pelton. It was set in Baskerville by Lone Wolf Enterprises, Ltd. The art director for the cover was Anthony Landi.

Printed and bound by Quebecor/Dubuque.

This book is printed on acid-free paper.

McGraw-Hill books are available at special quantity discounts to use as premiums and sales promotions, or for use in corporate training programs. For more information, please write to the Director of Special Sales, McGraw-Hill Professional, Two Penn Plaza, New York, NY 10121-2298. Or contact your local bookstore.

*This book is dedicated to
Adam and Afton,
the two brightest stars in my life.*

Introduction

Passing the licensing exam for a plumber's license is a major step toward a potentially lucrative career. A master plumber who goes into business can make a lot of money. Even working for an employer can give a plumber a better-than-average wage. There is money in plumbing, and it's usually not too difficult to find work. But, one of the first steps is passing the licensing exam. Many potential plumbers fail their licensing exam a time or two before they pass it. But, this book can help you be one of the people in the exam room who walk away with a passing grade and a plumber's license.

I've been in the plumbing trade for over 25 years. It was 1979 when I went into business for myself. I am a master plumber and a master gasfitter. During my career I have taught as adjunct faculty for Central Maine Technical College. One of the classes I taught was a code preparation class. This course was designed to help people pass their licensing exams. While teaching that class, I learned a lot about what potential plumbers have trouble with in studying for a licensing exam. While I can't look you in the eye and teach you one on one, I can help you with this book. My past students have enjoyed a high success rate in passing their exams. There is no reason why you should be left behind, now that this book is available to you.

Why do you need a study guide? The code can be difficult to understand and to interpret. Wording on exams can be very tricky. While you might be well versed in the code for field work, it would not be unusual for you to have difficulty passing an exam on the code regulations. Some people simply have trouble taking tests. The sample tests here will give you a feel for what to expect from your licensing exam. After taking these tests and passing them, you will have more confidence on the day when you take your real exam. There will be no need to freeze up when your exam is placed in front of you. Since you will have tested your knowledge here and studied your local code book, you will be well prepared for the licensing exam.

There is more here than simply a collection of test questions and answers. Take a moment to thumb through the pages. Notice all of the illustrations. They aren't pretty pictures used to make a bigger book. The illustrations are a type of nuts-and-bolts help for your study needs and your field work. Check out the conversion tables and think about how much time they can save you. Notice the diagrams for various types of vents. The visual picture of the venting procedures will make it easier to install the vents in the field.

Do you hate math? If you do, you're not alone. Yet, math is a part of plumbing. But don't worry, the help you need is in your hands. You will find numerous tables that do the math for you. Where a table isn't practical, you will find detailed formulas to make the math easier to calculate.

What is a plumber's license worth to you? It should be worth a lot. Fortunately, you don't have to pay a lot for it. All you need is this book, your local code book, some good study habits, and the price of your licensing exam. Investing in yourself and your license may well prove to be one of the best deals you will ever make. Go for it.

Contents

Chapter 1
DEFINITIONS

When you sit for your plumbing exam, you will sometimes see questions that could have more than one right answer. In these situations, the test will ask you to choose the best answer. If the example I've just given you were such a question, the correct answer would be Schedule-40 plastic pipe. Even though either pipe is an approved material, plastic pipe is the better choice.

This chapter is going to test your knowledge of approved plumbing materials. Many of the questions will have only one right answer, but some may have more than one correct answer. In these cases, select the best available answer.

The first test you will be given is a multiple-choice exam. Read the questions carefully; some of them are worded to catch you off guard, just as questions on your real exam may be. After reading the question, evaluate the possible answers and make your selection.

When you have completed the multiple-choice questions, move on to the true-false test. Do this before checking your score on the multiple-choice exam. Again, read the questions carefully and look out for trick wording.

Once you have completed all of the test questions, turn to the back of the chapter to find the correct answers for the questions. Grade your test and see how you made out.

Real plumbing exams are timed, so don't linger too long on any one question. Move through the test as if you were sitting for that all-important master's license.

After grading your test, spend some time reviewing the questions you answered incorrectly. Figure out why you made the wrong choice and commit the rule to memory. There will be no second chance on the day of your actual exam. Unless you are willing to wait for a new test session, pay an additional testing fee, and work for less money than a licensed plumber makes, you must pass your exam on the first try. This book can help you prepare to do just that, but you have to put forth the effort to learn, and you must take these sample exams seriously. Now let's get on with the tests.

KNOW YOUR DEFINITIONS

The plumbing code is full of terms that are not used on a regular basis by average people. Understanding code definitions is only a matter of memory. If you have a good memory, you will not have any trouble learning the definitions for words used in the plumbing code. However, when you are ready to take your plumbing exam, the meanings of certain words may escape your mind. What's worse, the exam may be worded in such a way as to confuse or trick you.

Definitions are usually not a major part of a licensing exam, but they do account for part of your test. Since every wrong answer on the exam will count against you, it is necessary to know the terms used in the code and their definitions. Some of the words you will already know, and others you will have to study. Don't overlook this aspect of your test. The definitions can be the pivotal point in passing or failing your exam.

If you open a code book, you will find many definitions. While many of the words seem simple enough to understand, it is easy to make a mistake on the day of your exam. Pure memorization of the text in the code book may not be enough to get you through the test. It has been my experience that many plumbing exams are designed to test more than your memory skills. They are meant to determine your understanding of the code and definitions. This fact gets some people in trouble.

One of the students who attended my class on code preparation had a very good memory. He could read a page of definitions and practically recite the words verbatim. His skills in memory work were exceptional. However, even with his strong ability to remember what the code said, the individual was not so gifted at applying the code. This fact came to light when I gave the class the test on definitions.

As an instructor, I attempt to simulate real-world situations. In regard to the tests that I create, this means that the wording in the questions must occasionally be a little tricky. The right type of questions can quickly pinpoint the people who have memorized the code rather than learned it. This was the case with the student who thought he knew all the definitions by heart. His work on the mock exam showed him that he didn't know the intent of the code. He knew the code but not its application. Fortunately, this person learned about his faults in my classroom rather than during the real exam.

This chapter is going to do for you what I did for my student in class. It is going to test your understanding of the definitions. I won't bore you by duplicating the definitions as they are presented in the plumbing code. Instead, I'll ask you questions that will determine just how well you really understand the words and terms. Let's start this testing with a multiple-choice exam. The questions in the test may have more than one answer that could be considered correct. Choose the answer that is most closely related to the code definitions.

MULTIPLE-CHOICE EXAM

1. When you read about types of pipe that may be used in a plumbing system in the code, you will often see the abbreviation ABS used to represent a particular type of pipe. What does ABS stand for:

   a. Anti-Block System
   b. Acrylonitrile-Butadiene-Styrene
   c. American-Butadiene-Styrene

2. When you have a shower valve in a wall that can be reached by removing an access panel, which of the following words best represents the condition of the shower valve:

   a. Accessible
   b. Readily accessible
   c. Partially accessible

3. If you walk into an open basement and observe the gate valve on a water heater, which of the following words best describes the condition of the gate valve:

   a. Accessible
   b. Readily accessible
   c. Partially accessible

4. When there is a physical separation between a discharge pipe and the drain that accepts the discharge, what do you have? Choose the word that best describes the situation:

   a. A broken pipe
   b. An airbreak
   c. An open drain

5. What term best describes the unobstructed vertical distance through the free atmosphere between the lowest opening from a pipe that is conveying waste to an open receptor:

   a. Free fall
   b. Airbreak
   c. Airgap

6. What term best describes the unobstructed vertical distance through the free atmosphere between the lowest opening from a pipe that is conveying potable water to the flood-level rim of a fixture:

   a. Free fall   b. Airbreak

   c. Airgap

7. When something is accepted or acceptable under an applicable standard of the plumbing code, it is said to be:

   a. All right   b. Compliant

   c. Approved

8. An organization that is primarily established for purposes of testing standards and whose existence has been approved by the proper authorities is known as:

   a. An engineering firm   b. An approved testing agency

   c. A certified plumbing company

9. When water enters a potable water pipe from any source other than its intended source, the process is known as:

   a. Reverse flow   b. Backflow

   c. Cross connection

10. What type of device should be installed in certain pipes to prevent the reverse flow of drainage substances:

   a. Reverse-flow valve   b. Backwater preventer

   c. Gate valve

11. Which of the following is the definition of a bathroom group:

   a. A group of fixtures consisting of a water closet, a bathing unit, and a lavatory, which may or may not include a bidet, an emergency floor drain, or both and that is located on the same floor level.

   b. A group of fixtures consisting of a water closet and a lavatory, which may or may not include a bidet, an emergency floor drain, or both and that is located on the same floor level.

   c. A group of water closets located in the same room.

12. What is the device that is present on a water heater that permits the emptying or discharge of excessive pressure:

   a. Relief valve   b. Boiler blowoff

   c. Aquastat

13. A connection or condition that may result in backflow is known as:

   a. A backflow connection   b. A backflow condition

   c. A backflow condenser

> **❗ Code**alert
>
> According to the 2006 Code, "approved" is now defined as anything accept-able to the code official or other authority having jurisdiction.

14. Increased pressure in a system can result in backflow. When backflow is caused by a pressure that is in excess of the supply pressure, the condition is known as:

    a. Back-puffing
    b. Back-pressure backflow
    c. Pressure imbalance

15. A device that is used to prevent backflow into a potable water supply is known as:

    a. Backwater valve
    b. Backwater preventer
    c. Backflow preventer

16. When used, contaminated, or polluted water from a plumbing fixture flows back into the potable water supply due to a negative pressure, the condition might be called backflow or:

    a. Reverse flo
    b. Back-siphonage
    c. Counter-siphonage

17. Any part of a piping system that is not a main, riser, or stack is a(n):

    a. Arterial
    b. Branch
    c. Feed

18. A vent that connects one or more individual vents with a vent stack or stack vent is a:

    a. Common vent
    b. Wet vent
    c. Branch vent

19. The lowest pipe in the drainage system, which receives the discharge from soil, waste, and other pipes within the inside walls of a building and conveys it to a pipe that begins 2 feet outside the building wall, is known as a:

    a. Sewer
    b. Building sewer
    c. Building drain

20. What pipe accepts the discharge from the pipe described in question 19:

    a. Sewer
    b. Building sewer
    c. Building drain

21.   What is another name for a water service?

    a.   Water main   b.   Building supply

    c.   Main inlet

22.   A lined excavation in the ground that receives the discharge of a drainage system that has been designed to retain organic matter and solids while allowing liquids to seep through the bottom and sides is known as a:

    a.   Leach field   b.   Septic tank

    c.   Cesspool

23.   A plumbing system that is designed with the waste piping embodying the horizontal wet venting of one or more sinks or floor drains by means of a common waste and vent pipe that is adequately sized to provide free movement of air above the flow line of the drain is known as:

    a.   A Combination waste and vent system

    b.   An engineered system

    c.   An illegal system

24.   An impairment of the quality of potable water that creates a hazard to public health through poisoning or the spread of disease is known as:

    a.   Contamination   b.   Backflow

    c.   Risk danger

25.   A vent that is run vertically and extends as a continuation of the drain it serves is known as a:

    a.   Common vent   b.   Vent stack

    c.   Continuous vent

26.   A waste arrangement that connects multiple compartments of a set of fixtures, such as a double-bowl kitchen sink, to a single trap is known as:

    a.   An island waste   b.   A collective waste

    c.   A continuous waste

27.   A connection between a potable water supply and any fixture or device where non-potable water may enter the potable supply is known as:

    a.   A cross connection   b.   A reverse connection

    c.   A backflow connection

28.   The minimum elevation above the flood level rim of a fixture or receptacle that a a backflow preventer can be installed is known as the:

    a.   Cutoff point   b.   Critical level

    c.   Installation level

29. When a pipe is measured along the center line and fittings used with the pipe, the measurement determines the:

   a. Combined length   b. Developed length

   c. Overall length

30. A pipe that conveys waste or water-borne wastes in a building drainage system is known as:

   a. A branch   b. A stack

   c. A drain

31. Liquid and water-borne wastes that are derived from ordinary living processes and are free of industrial or other wastes that may require special treatment other than the simple discharging into a public sewer is known as:

   a. Common sewage   b. Safe sewage

   c. Domestic sewage

32. A plumbing system or any part of a system that was installed prior to the effective date of the plumbing code is deemed to be:

   a. Approved work   b. Existing work

   c. Exempt work

33. A water-supply pipe between a fixture-supply pipe and a water-distribution pipe is known as a:

   a. Branch fixture   b. Fixture branch

   c. Fixture supply

34. The drainage piping that extends from the trap of a fixture to a point of connection with another drain pipe is called a:

   a. Trap arm   b. Horizontal stack

   c. Fixture drain

35. A water-supply pipe that connects a fixture with a fixture branch is called a:

   a. Durham fitting   b. Fixture supply

   c. Fixture feed

! Codealert

The new code defines a branch interval as a vertical measurement of distance, 8 feet (2438 mm) or more in developed length, between the connections of horizontal branches to a drainage stack. Measurements are taken down the stack from the highest horizontal branch connection.

36. The top edge of a fixture or receptacle from which water can overflow is called a:

    a. Critical level   b. Flood-Level rim

    c. Flood surface

37. The slope or fall of a drainage pipe is normally called either the pitch of the pipe or the _____ of the pipe. Fill in the blank:

    a. Elevation   b. Drop

    c. Grade

38. A device designed to retain grease from up to four fixtures is called a:

    a. Grease trap   b. Grease interceptor

    c. Grease diverter

39. A device with at least a 750-gallon capacity that is meant to serve one or more fixtures in preventing grease from entering the building drain and is remotely located is called a:

    a. Grease trap   b. Grease interceptor

    c. Grease diverter

40. Another name for a house drain is a:

    a. Building sewer   b. Building drain

    c. Building trap

41. Another name for a house sewer is:

    a. Building sewer   b. Building drain

    c. Building trap

42. A vertical pipe must be installed vertically and form no angles that are greater than _____ degrees with the vertical. Fill in the blank.

    a. 22 1/2   b. 45

    c. 60

43. A pipe that is installed to vent a fixture trap and that connects with the vent system above the fixture or terminates into the open air is called a(n):

    a. Individual vent   b. Vacuum vent

    c. Wet vent

44. Any liquid or water-borne waste from industrial or commercial processes other than domestic sewage is known as:

    a. Industrial waste   b. Commercial waste

    c. Either A or B

45. A pipe that does not connect directly with the drainage system but conveys liquid wastes by discharging into a plumbing fixture connected to the drainage system is called an:

    a. Airgap   b. Open-air waste

    c. Indirect waste

46. When a condition contrary to sanitary principles or injurious to health exists, that condition is known as:

    a. Unsanitary   b. Insanitary

    c. Detrimental

47. Discharge that does not contain fecal matter is called:

    a. Chemical waste   b. Liquid waste

    c. Modified sewage

48. When a plumbing fixture is not supplied with water sufficient to flush it and maintain a clean condition, the fixture is deemed to be:

    a. Sanitary   b. Insanitary

    c. Detrimental

49. Any trap that does not maintain a proper seal is deemed to be:

    a. Useless   b. Insanitary

    c. Detrimental

50. A vent that connects a horizontal branch or fixture drain with a stack vent of the originating waste or soil stack is called a:

    a. Relief vent   b. Bombay vent

    c. Loop vent

51. The principal artery of a venting system to which vent branches may be connected is called a:

    a. Vent stack   b. Stack vent

    c. Main vent

52. A combination of elbows or bends that brings one section of pipe out of line but into a line parallel with the other section of the pipe is known as:

    a. An offset   b. A drainage turn

    c. A crown vent

53. What is the abbreviation for polybutylene:

    a. PB   b. PE

    c. PBL

54. What is the abbreviation for polyethylene:

    a. CPVC   b. PE

    c. PEL

55. Potable water is water that may be used for which of the following:

    a. Domestic purposes   b. Culinary purposes

    c. Both A and B

56. Another name for drinking water is:

    a. Potable water   b. Gray water

    c. Chlorinated water

57. A sewer that receives the discharge from more than one building drain and conveys it to a point of disposal is called a:

    a. Common sewer   b. Elevated sewer

    c. Private sewer

58. The letters PVC are an abbreviation for:

    a. Polyvinyl chloride   b. Polyvinyl clusters

    c. Polyvinyl calcium

59. A plumbing fixture or device that has been designed and approved to receive the discharge from indirect waste pipes is called:

    a. A dump station   b. A receptor

    c. An airbreak

60. A vent whose primary function is to provide circulation of air between drainage and vent systems or to act as an auxiliary vent on a specially designed system is known as a:

    a. Relief vent   b. Loop vent

    c. Vent stack

> **! Code**alert
>
> **FLOW CONTROL (VENTED):** A device installed upstream from the interceptor having an orifice that controls the rate of flow through the interceptor and an air intake (vent) downstream from the orifice that allows air to be drawn into the flow stream.

61.   The unobstructed open edge of a fixture is known as the:

   a.   Rim   b.   Overflow

   c.   Edge

62.   A lined excavation in the ground that receives the discharge of a septic tank so designed as to permit the effluent from the tank to seep through its bottom and sides is known as a:

   a.   Leach field   b.   Cesspool

   c.   Seepage pit

63.   Any liquid waste that contains animal or vegetable matter in suspension or solution is known as:

   a.   Liquid waste   b.   Sewage

   c.   Domestic waste

64.   An adjustable tubing connection that consists of a compression nut, friction ring, and compression washer designed to fit a threaded adapter fitting or a standard taper pipe thread is called a:

   a.   Slip joint   b.   Temporary connection

   c.   Compression joint

65.   The vertical main of a system of soil, waste, or vent piping that extends through one of more stories is called a:

   a.   Stack   b.   Riser

   c.   Stack vent

66.   A soldered joint is one obtained by joining metal parts with metallic mixtures or alloys that melt at a temperature up to and including _____ degrees F. Fill in the blank:

   a.   840 degrees   b.   875 degrees

   c.   925 degrees

67.   A water-supply pipe that extends vertically for one full story or more, is called a:

   a.   Stack   b.   Riser

   c.   Main

68. Wastes that require a special method of handling are known as:
    a. Chemical wastes   b. Special wastes
    c. Industrial wastes

69. A pipe that is an extension of a soil or waste stack above the highest horizontal drain is called a:
    a. Stack vent   b. Vent stack
    c. Main stack

70. A pipe that connects upward from a soil or waste stack to a vent stack to prevent pressure changes in the stacks is known as:
    a. A relief vent   b. A pressure-reducing vent
    c. A yoke vent

71. An approved tank or pit that collects sewage or liquid waste, is located below the normal grade of the gravity plumbing system, and requires mechanical means to be emptied is called a:
    a. Cesspool   b. Sewage well
    c. Sump

72. The portion of a fixture drain that runs between the trap of the fixture and the vent for the fixture is known as a:
    a. Fixture drain   b. Trap arm
    c. Wet vent

73. The maximum vertical depth of liquid that a trap will retain, measured between the crown weir and the top of the dip of the trap, is known as:
    a. The trap seal   b. The trap level
    c. The critical level

74. A vertical vent pipe installed primarily to provide circulation of air to and from any part of the drainage system is known as a:
    a. Vent stack   b. Stack vent
    c. Common vent

75. A water well that is made by boring a hole into the ground with an auger and installing a casing is a:
    a. Driven well   b. Drilled well
    c. Bored well

76. ABS is the acronym for which of the following?
    a. Automated Bank Systems   b. Acrylonitrile-Butadiene-Styrene
    c. Approved Branch System

> **! Code**alert
>
> **GREASE INTERCEPTOR:** A plumbing appurtenance that is installed in a sanitary drainage system to intercept oily and greasy wastes from a wastewater discharge. Such device has the ability to intercept free-floating fats and oils.

77. When a device is deemed _____, it is within code requirements for it to be concealed by a removal panel or plate.

   a. Accessible

   b. Hard to reach

   c. Readily Accessible

78. To collect storm or surface water from an open area, you need to install which of the following:

   a. A water filter

   b. An air gap

   c. An area drain

79. Backsiphonage is best described as:

   a. Discharged water

   b. Contaminated backflow

   c. Backpressure

80. A ballcock is usually found in which of the following:

   a. Toilet tanks

   b. Boilers

   c. Branch Fixtures

81. A Branch can be considered any part of a piping system that is not a riser, a main, or a _____.

   a. Vent

   b. Branch Interval

   c. Stack

82. A joint made by joining metal parts with alloys that melt at temperatures higher than 840 degrees F, but lower than the melting point of the parts being joined, is called:

   a. Very hot

   b. Soldered

   c. Both a and b

83. A grease trap has a passive interceptor device whose rated flow is _____ GPM or less.

   a. 25

   b. 100

   c. 50

84. Hot water is considered to have a temperature greater than or equal to this degree:

   a. 150 degrees

   b. 120 degrees

   c. 100 degrees

> **! Code**alert
>
> **GREASE REMOVAL DEVICE, AUTOMATIC GRID:** A plumbing ap
> purtenance that is installed in the sanitary drainage system to intercept free-
> floating fats, oils, and grease from wastewater drainage. Such a device oper
> ates on a time- or event-controlled basis and has the ability to remove
> free-floating fats, oils, and grease automatically without intervention from the
> user, except for maintenance.

85. The lowest portion of the inside of a horizontal pipe is called what?

   a. Invert   b. Extrovert

   c. Interceptor

Lay your pencil down and rest your eyes. You've just completed the multiple-choice exam on definitions. Don't check your answers yet; there is still a true-false test to take.

How do you think you did on this part of the test? Did some of the questions pose problems for you? Unless you possess an uncommon grasp of code definitions, you probably were confused by some of the potential answers. Don't feel bad; you were supposed to be. The purpose of this book is to make you think and to expose your weak areas. By doing this, you will be better prepared for the real exam, when the score will determine whether you get your plumber's license or not. Take a short break if you like, and then move on to the true-false questions.

TRUE-FALSE EXAM

1. A pipe that conveys potable water from a building supply to a plumbing fixture is a water-distribution pipe.

True   False

2. A welded joint is any joint obtained by the joining of metal parts in the plastic molten state.

True   False

3. A waste pipe is a pipe that conveys liquid waste and suspended fecal matter.

True   False

4. The purpose of a trap is to prevent the back passage of air without materially affecting the flow of sewage or waste water through it.

True   False

5. Wet vents also serve as drains.

True   False

6. A soil pipe is a pipe that conveys the waste of sump pumps, laundry tubs, and lavatories.

True   False

7. A sewer that is directly controlled by a public authority and that serves as a common sewer is known as a public sewer.

True   False

8. Inadequate or unsafe water-supply or sewage-disposal systems are considered to be a nuisance.

True   False

9. A dug well is one that is made by excavating a large-diameter shaft and installing a casing.

True   False

10. Grease interceptor is just another name for a grease trap.

True   False

11. Nonpotable water is safe to drink.

True   False

12. The amount of pressure present when there is no flow is called Static Pressure.

True   False

13. A relief vent is any vent that provides air circulation between drainage and vent systems.

True   False

14. You never need to have a code officer's approval as long as the homeowner agrees to the work being done.

True   False

15. A pipe or tube that connects the outlet of a plumbing fixture to a trap is called a vacuum vent.

True   False

16. A wet vent serves only as a vent.

True   False

17. A gang or group shower is two or more showers in a common area.

True   False

18. SDR refers to Standard dimensional ratio.

True   False

19.   Plumbing codes are not necessary in all states.

       True   False

20.   A vertical pipe does not make an angle of more than 45 degrees.

       True   False

21.   Access refers to some means of making devices reachable.

       True   False

22.   An adapter fitting is any approved fitting that can be used to connect pipes and fittings that would other-wise not fit together.

       True   False

23.   A cistern is a storage tank that is used to collect and store storm water for uses associated with potable water.

       True   False

24.   Any structure containing building materials that will ignite and burn at temperature of 1392 degrees F. or less is combustible construction.

       True   False

25.   Existing work and existing installations have the same meaning.

       True   False

MULTIPLE-CHOICE ANSWERS

1.	b	18.	c	35.	b	52.	a	69.	b
2.	a	19.	b	36.	b	53.	a	70.	c
3.	b	20.	a	37.	c	54.	b	71.	c
4.	b	21.	b	38.	a	55.	c	72.	b
5.	c	22.	c	39.	b	56.	a	73.	a
6.	c	23.	a	40.	b	57.	a	74.	a
7.	c	24.	a	41.	a	58.	a	75.	b
8.	b	25.	c	42.	b	59.	b	76.	b
9.	b	26.	c	43.	a	60.	a	77.	a
10.	b	27.	a	44.	a	61.	a	78.	c
11.	a	28.	b	45.	c	62.	a	79.	b
12.	a	29.	b	46.	b	63.	b	80.	a
13.	a	30.	c	47.	b	64.	a	81.	c
14.	b	31.	c	48.	b	65.	a	82.	b
15.	c	32.	b	49.	b	66.	a	83.	c
16.	b	33.	b	50.	a	67.	b	84.	b
17.	b	34.	a	51.	c	68.	b	85.	a

TRUE-FALSE ANSWERS

1.	True	8.	True	14.	False	20.	True
2.	True	9.	True	15.	False	21.	True
3.	True	10.	False	16.	False	22.	True
4.	True	11.	False	17.	True	23.	False
5.	True	12.	True	18.	True	24.	True
6.	False	13.	True	19.	False	25.	True
7.	True						

Chapter 2
ADMINISTRATIVE POLICIES AND PROCEDURES

Administrative and enforcement issues pertain to both journeyman and master plumbers. The rules for master plumbers are a bit more complex and comprehensive than they are for journeyman plumbers. This does not, however, mean that journeyman plumbers have no responsibilities under the rules pertaining to administration and enforcement. In fact, it is this very section of the code book that relates to the need to be properly licensed in order to perform plumbing work.

When you open a plumbing code book, the first part of the book deals with administration. Why is this? Because it is the administration of the code that makes the rest of the code requirements enforceable. To work within the parameters of a plumbing code, you must understand how the code is administered and enforced. This chapter is going to test your knowledge in the areas of administration and enforcement.

Will there be questions about administration and enforcement on the exam for your license? It's a safe bet to say there will be. Most plumbing exams are altered to some extent on a rotating basis to keep people from cheating, but issues of administration and enforcement are common topics found in all plumbing exams. How many of the questions on the exam will be derived from the administration section of the code? This question has no uniform answer. Each plumbing exam can be different. The only way to be properly prepared for such an exam is to know all the material that may be presented on the test.

As a code instructor, I have taught a lot of people how to interpret the plumbing code and how to prepare for a licensing exam. Some of the people have been apprentices, and some have been journeyman plumbers. The journeyman plumbers were preparing for the exam to obtain a master's license. It is amazing how few licensed plumbers have a good working knowledge of administration and enforcement issues. While most journeyman plumbers know how far a vent may be from a drain and what the minimum size of a water service must be, darn few of them can answer a handful of questions correctly on administration issues.

I've trained a lot of plumbers, both as a master plumber and as a code instructor, and I'm sure that a majority of them had little respect for the administration section of a code book. It's not that they didn't intend to comply with the plumbing code; it is just that they didn't seem to believe that the administration aspect of the code was important to learn for their exams.

When you sit for your exam, you are going to be allowed to answer a certain percentage of the questions incorrectly. If you exceed the percentage of incorrect answers allowed, you will not pass the test. A question about administration is just as important as a question about chemical wastes when it comes time to tally up the right and wrong answers. A miss is a miss, and a failing score is the same no matter why you fail. I tell you this in hope that you won't underestimate the administration and enforcement section of your exam.

Since master plumbers usually deal with administration issues, many prospective journeyman plumbers don't feel the administration section of the code applies to them. This simply isn't true. Yes, the section is more pertinent to master plumbers, but it certainly does apply to journeyman plumbers as well.

Every local jurisdiction is allowed to adapt and alter a plumbing code. Changes made on the local level don't show up in a typical code book. Before you become completely comfortable with your grasp of the code, check with your local code-enforcement office to see if there are any specials rules or regulations required by the local jurisdiction that may be in conflict with what the code book states. Local changes are usually minor, and they are not normally going to be extensive enough to have an impact on your exam results. But you will be wise to confirm this fact before taking your real test. Now, let's get on with the sample tests.

MULTIPLE CHOICE EXAM

1.   When a matter that is essential for sanitary safety pertaining to an existing building is not covered in the code book, who has the authority to make a decision on how the matter should be handled:

    a. A master   b. A journeyman plumber

    c. A code official   d. None of the above

2.   When a matter that is essential for sanitary safety pertaining to a new building is not covered in the code book, who has the authority to make a decision on how the matter should be handled:

    a. A master plumber   b. A journeyman plumber

    c. A code official   d. None of the above

3.   If a part of a building is in violation of the code and a code officer condemns the building, may the building continue to be used without violating the code:

    a. Yes

    b. No

    c. Yes, so long as a written agreement is entered into

    d. Yes, so long as the owner of the property agrees to correct the code violation within 14 days

4.   Which of the following aspects of plumbing systems are governed by the plumbing code:

    a. Sanitary drainage   b. Storm drainage

    c. Sanitary facilities   d. All of the above

5.   Which of the following aspects of plumbing systems are governed by the plumbing code:

    a. Water supplies   b. Storm water

    c. Sewage disposal   d. All of the above

6.   Which of the following is the intent of the plumbing code:

    a. To produce revenue for the local jurisdiction

    b. To ensure public safety, health, and welfare

    c. To regulate plumbers

    d. All of the above

7.   When is an alteration or repair to an existing plumbing system in violation of the plumbing code:

    a. When it is done without a permit

    b. When it is not done by a master plumber

    c. When it renders the plumbing system unsafe

    d. Any of the above

8. When is an alteration or repair to an existing plumbing system in violation of the plumbing code:

    a. When it is done without prior code approval

    b. When it is not done by a master plumber

    c. When it adversely affects the performance of the plumbing system

    d. Any of the above

9. This question calls for you to assume a certain scenario and make a decision. It is your job to add a bathroom in a home where two full bathrooms presently exist. The building drain and sewer for the house are both 3-inch pipe. Current code regulations limit the installation of water closets on a 3-inch pipe to no more than two toilets. However, you are working with existing plumbing, so do you have to change the size of the existing pipes or can you connect to them with the new toilet? The plumbing code allows some exceptions from current rules for existing plumbing that is being altered. If you found yourself in this position, what would you do? Your decision and ruling in the code book may not be identical, so answer this question with the correct answer as it appears in the code:

    a. Consult the local code officer for a ruling

    b. Connect to the existing 3-inch pipe with the new toilet, since the existing pipe is grandfathered in the code

    c. Increase the size of the building drain or sewer from the point of connection that exceeds the limit of two water closets

    d. Ask your boss what to do

10. This question also puts you into a specific situation. The owner of a building that was used previously as a retail store has requested that the plumbing be converted to accommodate a hair salon. What are you required by code to do? Answer this question with the best possible answer; more than one of the answer may apply, but only one is the best answer:

    a. Gain approval and certification for the proposed changes

    b. Install proper backflow-prevention devices

    c. Defer to a master plumber

    d. Use materials approved by the local code-enforcement office

11. All plumbing repairs and replacements are required to be made in what manner:

    a. By a master plumber

    b. In compliance with the local plumbing code

    c. By a journeyman plumber

    d. Both A and B

12. All plumbing repairs and replacements are required to be made in what manner:

    a. By a master plumber

    b. In compliance with the local plumbing code

    c. In a safe and sanitary manner

    d. Both B and C

13. When minor repairs or replacements are made to an existing plumbing system, which of the following rules apply:

    a. All work must be brought up to current code requirements for new plumbing

    b. The repair or replacement may be made in the same manner and arrangement as in the existing system

    c. A permit must be obtained

    d. Both B and C

14. All plumbing systems, both new and existing, must be maintained to assure which of the following:

    a. That the system is upgraded periodically to comply with changes in the plumbing code

    b. That the system remains in good working order and continues to be both safe and sanitary

    c. That no unlicensed individuals alter the system in any way

    d. All of the above

15. Who is responsible for the maintenance of a plumbing system:

    a. A master plumber

    b. The property owner

    c. The property owner's designated agent

    d. Either B or C

16. If a building is scheduled to be demolished, who is responsible for making proper notifications to the people or agencies who need to be informed:

    a. A master plumber

    b. The property owner

    c. The property owner's designated agent

    d. Either B or C

17. If a building is scheduled to be demolished, who is to be notified of the intended work:

    a. A master plumber

    b. The property owner

    c. The property owner's designated agent

    d. The principals of utilities with service connections involving the building

18. Who has a right to modify the provisions of the local plumbing code:

    a. The mayor

    b. A master plumber

    c. A code official

    d. Any of the above

19. Variances to the code are sometimes approved if it is deemed by the proper authorities that there is just cause for the request. If a variance is requested, who is required to make the request:

  a. A master plumber

  b. A property owner

  c. A property owner's representative

  d. A journeyman plumber

  e. Either A or D

  f. Either B or C

20. When a variance to a code requirement is requested, how should the request be made:

  a. In person

  b. By phone

  c. In writing

  d. In the presence of the appeal board

21. When a variance from the plumbing code is requested, it may be approved if the alteration is considered to be in the spirit and intent of the code and does not represent a risk to public health, safety, and welfare. Who has the authority to grant a variance:

  a. The clerk of the court

  b. The mayor

  c. A code official

  d. An independent engineering firm

22. Who can authorize the use of previously used plumbing materials and equipment:

  a. A master plumber

  b. A property owner

  c. A property owner's representative

  d. A local code official

23. Before used plumbing materials or equipment can be approved for use in new applications, the materials or equipment must be:

  a. Reconditioned

  b. Tested

  c. Placed in working order

  d. Approved

  e. All of the above

24.   Who has the authority to allow the use of materials and equipment that are not covered in the code book?

   a.   A property owner

   b.   A property owner's representative

   c.   An independent engineering firm

   d.   A code official

25.   For an alternative material or piece of equipment to be approved for use, it must meet which of the following criteria:

   a.   The design must be satisfactory

   b.   It must comply with the intent of the code

   c.   It must be approved by the property owner

   d.   Both A and B

   e.   Both B and C

   f.   It must be approved by the Master Plumber's Guild

26.   For an alternative type of material or equipment to be approved for use, it will be considered by which of the following criteria:

   a.   Quality   b.   Strength

   c.   Effectiveness   d.   Fire resistance

   e.   Durability   f.   Safety

   g.   All of the above   h.   Both A and E

27.   If an alternative material is submitted for approval, the local code official may request that research and investigation be conducted to prove the material suitable for use. Who is responsible for the cost of such research and investigation:

   a.   The master plumber

   b.   The property owner

   c.   The applicant of the request for acceptance

   d.   The code-enforcement office

28.   According to the plumbing code, who is the executive official in charge of the code:

   a.   The code official   b.   The clerk of the court

   c.   The mayor   d.   The town council

29.   Who is responsible for code-compliance inspections of plumbing systems:

   a.   The master plumber   b.   The property owner

   c.   The code official   d.   The fire marshal

30. Who is responsible for keeping records of applications received, permits and certificates issued, fees collected, reports of inspections made, and notices and orders issued in regard to the plumbing code:

   a. The health department

   b. The public-works department

   c. The local code official

   d. The master plumber involved in the work being performed

31. When a plumbing permit is required for proposed work, the work may not be commenced until:

   a. A permit is applied for

   b. A permit is issued

   c. A permit is approved

   d. A permit is signed by a code official

32. Which of the following types of work requires a plumbing permit:

   a. The replacement of an electric water heater

   b. The repair of a pipe that froze and split

   c. The replacement of a faucet stem

   d. The clearance of a drain stoppage

   e. The replacement of a faulty gate valve

33. Which of the following individuals will not normally be eligible to receive a plumbing permit?

   a. A master plumber

   b. A journeyman plumber

   c. A property owner who will personally perform all plumbing work described in the permit application on a primary residence owned by that person

   d. Both B and C

34. A plumbing permit is not valid until it is:

   a. Posted in a conspicuous place on the job site

   b. Signed by the master plumber or an authorized representative

   c. Signed by the code official or an authorized representative

   d. Both B and C

35. As a journeyman plumber, you will not bear the burden that a master plumber does in the areas of administration that pertain to permit acquisition. You will, however, be responsible for parts of the compliance required by a permit. As a journeyman plumber who is installing plumbing under a permit obtained by a master plumber, you must adhere to which of the following rules:

    a. All work must be done under the direct supervision of a master plumber

    b. All work must be done in compliance with the plumbing code

    c. All work must be done in the manner described to the code official at the time of application for the permit

    d. All of the above

    e. Both A and B

    f. Both B and C

36. As a journeyman plumber working on a job when a stop-work order is issued by a code officer, what must you immediately do:

    a. Inform the master plumber

    b. Quit working on the job

    c. Request a written explanation for the order

    d. Finish the day and deliver the notice to your master plumber

37. While you are working on a job as a journeyman plumber, you are required to have which records accessible to the local code official at all times?

    a. Material invoices

    b. Test results from previous inspections

    c. Previous inspection results

    d. Both B and C

38. As a journeyman plumber, you are required to perform all plumbing work in which of the following ways:

    a. A workmanlike manner

    b. An acceptable manner

    c. Under the direction of the property owner

    d. All of the above

    e. Both A and B

    f. Both A and C

39. When will a code officer issue a notice of approval for a plumbing system:

    a. When all work is complete

    b. When all tests and inspections prove the work is in compliance with the local plumbing code

    c. When requested, in writing, to do so

    d. When the owner of a property requests a notice of approval

40.   According to the plumbing code, what constitutes good workmanship:

    a.   Any work done by a licensed plumber

    b.   Any work that is clean and neat

    c.   Any work that secures the results intended by the plumbing code

    d.   Any work acceptable to the property owner

41.   Work that requires a plumbing permit may not be started until _____. Fill in the blank:

    a.   A permit is applied for

    b.   A permit is issued

    c.   Either A or B

    d.   Neither A or B

42.   A permit is not required for repairs that involve only _____. Fill in the blank:

    a.   The relocation of fixtures

    b.   The working parts of faucets and valves

    c.   The replacement of water heaters

    d.   None of the above

43.   A permit is not required for repairs that involve only _____. Fill in the blank:

    a.   The clearing of stoppages

    b.   The working parts of faucets and valves

    c.   Both A and B

    d.   None of the above

44.   A permit is not required for repairs that involve only _____. Fill in the blank:

    a.   The clearing of stoppages

    b.   The working parts of faucets and valves

    c.   The replacement of defective valves or faucets

    d.   All of the above

45.   Which of the following persons are allowed to apply for a plumbing permit?

    a.   Journeyman plumbers   b.   Master plumbers

    c.   The general public   d.   None of the above

46.   Which of the following persons are allowed to apply for a plumbing permit?

    a.   Registered design professionals   b.   Master plumbers

    c.   Property owners   d.   Any of the above

47. Which of the following persons are allowed to apply for a plumbing permit:

    a. Lessees of buildings

    b. Master plumbers

    c. Agents of property owners

    d. Any of the above

48. Which of the following are key elements that must be included in an application for a plumbing permit:

    a. The number of fixtures to be installed

    b. The location of the work

    c. The occupancy of the building in which the work is to be done

    d. All of the above

49. How many sets of construction documents must be provided during the application for a plumbing permit:

    a. 1

    b. 2

    c. 3

    d. 4

50. Which of the following elements is a substantial part of the construction documents required in order to obtain a plumbing permit:

    a. Floor plans

    b. Riser diagrams

    c. Both A and B

    d. Neither A or B

51. Construction documents presented with a permit application may be required to provide information on _____. Fill in the blank:

    a. The direction of flow

    b. The quantity of flow at a specific time of the day

    c. The individual plumber who will install the plumbing system

    d. None of the above

52. Construction documents presented with a permit application may be required to provide information on _____. Fill in the blank:

    a. The grade of horizontal piping

    b. The elevations relevant to the proposed plumbing

    c. Both A and B

    d. None of the above

53. Construction documents presented with a permit application may be required to provide information on _____. Fill in the blank:

    a. The drainage fixture-unit load on the system

    b. The supply fixture-unit load on the system

    c. Both A and B

    d. None of the above

54. A site plan will normally show the location of _____. Fill in the blank:

    a. The proposed water service   b. The proposed sewer connection

    c. Both A and B   d. None of the above

55. Vent-stack terminations should be shown on a site plan with respect to _____. Fill in the blank:

    a. The number of plumbing fixtures installed

    b. The type of plumbing fixtures installed

    c. Building ventilation openings

    d. None of the above

56. A permit application will be considered abandoned if the application is not diligently prosecuted or a permit issued within a _____ month period of time. Fill in the blank:

    a. 3   b. 6

    c. 9   d. 12

57. A code official may extend the time allowed for a permit application by a period of _____ days. Fill in the blank:

    a. 30   b. 60

    c. 90   d. 120

    e. 180

58. Any plumbing permit issued may become invalid if work is not started within _____ months from the date the permit was issued. Fill in the blank:

    a. 3   b. 6

    c. 9   d. 12

59. Any plumbing permit issued may become invalid if work ceases for a period of time that exceeds _____ months. Fill in the blank:

    a. 3   b. 6

    c. 9   d. 12

60. All approved construction documents will be _____. Fill in the blank:

    a. Stamped by a code official   b. Endorsed by a code official

    c. Both A and B   d. Either A or B

61. Which of the following is likely to be located on a site plan:

    a. A riser diagram

    b. A stack diagram

    c. The location of a private sewage disposal system if a public sewer is not available

    d. All of the above

62.   Which of the following locations is required to be supplied with a set of approved construction documents?

    a.   The code-enforcement office   b.   The job site

    c.   Both A and B   d.   Either A or B

63.   If a false statement is made during the application for a permit, what is likely to happen?

    a.   The permit will be suspended

    b.   The permit will be revoked

    c.   The master plumber will go to jail

    d.   The property owner will be sued

64.   Prior to the approval of construction documents by a code official, which of the following may be allowed?

    a.   All plumbing work may be installed, so long as a permit is applied for within 90 days of the completion of work

    b.   A code officer can authorize a partial approval, allowing some of the work to be done

    c.   A master plumber must apply to a board of appeals

    d.   None of the above

65.   How much notice is normally required to be given to a code official before a job inspection can be expected:

    a.   8 hours   b.   24 hours

    c.   48 hours   d.   72 hours

66.   Where must a true copy of a permit be kept during the entire time of work being done under that permit:

    a.   Displayed in the master plumber's office

    b.   At the code-enforcement office

    c.   On the job site

    d.   Any of the above

67.   Under what conditions does a code officer have the authority to order and require occupants of a building to vacate a structure forthwith:

    a.   Code officers do not posses such a power

    b.   When a senior code officer approves the action

    c.   When in the opinion of the code officer imminent danger of contamination or a sanitation hazard may endanger life

    d.   When a code violation is discovered

68.   Which of the following may request an appeal from a board of appeals:

    a.   A permit holder   b.   A property owner

    c.   A master plumber   d.   Any of the above

69.   Which of the following is a suitable cause for requesting an appeal:

    a.   A dislike for the present plumbing code

    b.   A denied permit application

    c.   A claim that the true intent of the code has been interpreted incorrectly

    d.   Any of the above

70.   Which of the following is a suitable cause for requesting an appeal:

    a.   An opinion that the provisions of the code do not apply fully

    b.   A denied permit application

    c.   A claim that the true intent of the code has been interpreted incorrectly

    d.   Both A and C

71.   Which of the following is a suitable cause for requesting an appeal?

    a.   An opinion that the provisions of the code do not apply fully

    b.   When an equally good or better form of construction is proposed

    c.   A claim that the true intent of the code has been interpreted incorrectly

    d.   Any of the above

72.   Where are forms for an appeal obtained?

    a.   From the appeals board   b.   From a code officer

    c.   From a post office   d.   Any of the above

73.   How many days are allowed to file for an appeal after notice has been served:

    a.   3   b.   5

    c.   10   d.   20

74.   How many people are required to form a board of appeals:

    a.   3   b.   5

    c.   7   d.   9

75. Under what conditions is a member of the board of appeals unable to hear a case:
    a. When the member has a personal interest in the case
    b. When the member has a professional interest in the case
    c. When the member has a financial interest in the case
    d. Any of the above

76. Potable water supplies fall under the jurisdiction of:
    a. building codes
    b. plumbing codes
    c. your local town office
    d. town water supply

77. When conditions warrant a hardship, a property owner or his agent, may request _____ from the standard code requirement.
    a. an amendment
    b. an extension of payment
    c. a variance
    d. an alternative method waiver

78. One aspect of a Code officers job is to:
    a. Maintain records
    b. Approve water heater venting systems
    c. Assist with installation
    d. Provide technical data for using alternate materials

79. The permit application shall include this information.
    a. A full description of the plumbing
    b. The location of the site
    c. Both a and b
    d. A set of detailed plans

80. Administrative policies dictate the procedure for code enforcement, interpretation and _____.
    a. mediation
    b. decision making
    c. plumbing
    d. implementation

81. You must always obtain a plumbing permit to replace a:
    a. Kitchen Sink
    b. Water Heater
    c. Sump pump
    d. Grease Trap

82. In the event that a structure is to be demolished or moved it is up to this person to notify all utility companies.
    a. The homeowner or agent
    b. The Code officer
    c. The Plumber
    d. The local town office

83.   Within date of issue, a plumbing permit is good for how long?

    a.   A plumbing permit doesn't expire.

    b.   One year

    c.   Six months

    d.   Four months

84.   A set of approved plans must be kept _____.

    a.   in a safe so they won't get lost   b.   at the Code office

    c.   in your glove compartment   d.   at the job site

85.   Code officers are responsible for the _____ and _____of the plumbing code.

    a.   approval and changes   b.   description and location

    c.   administration and enforcement d.   application and disbursement

We have reached the end of the multiple-choice exam. Don't turn to the back of the chapter to check the results of your answers at this time. Instead, proceed directly to the true-false questions. When you have completed all the test questions, you may turn to the answers and check your work.

TRUE-FALSE EXAM

1.   The plumbing code is in existence to provide minimum required standards for the protection of public health, safety and welfare.

    True   False

2.   The plumbing code applies to the erection, installation, alteration, repair, relocation, replacement, addition to, use, andr maintenance of plumbing systems.

    True   False

3.   Any alteration to an existing plumbing system requires that the present system be upgraded to current code requirements.

    True   False

4.   Master plumbers shall have the right to duly appoint and enforce the plumbing code.

    True   False

5.   A plumbing permit is not required to fix leaks in drains that are not concealed.

    True   False

6.   If a concealed trap or drain becomes defective and must be replaced, the job shall be considered as new work and a permit will be required.

     True   False

7.   The removal and reinstallation of a water closet does not require a permit so long as the work does not require the replacement or alteration of valves and pipes.

     True   False

8.   A set of approved plans and specifications that have been accepted by the code official must be kept in the office of the permit holder at all times.

     True   False

9.   The administrative authority may require a notice of at least one workday prior to making an on-site plumbing inspection.

     True   False

10.   It is a violation of the plumbing code to connect to a water-supply line or sewer without the permission of the administrative authority.

      True   False

11.   Unlicensed plumbers can work anywhere they want as long as they obtain a permit.

      True   False

12.   The Code officer may require a detailed set of plans before issuing permits.

      True   False

13.   It is the Code officer's job to inspect work after it is done to ensure that you didn't overcharge the customer.

      True   False

14.   A Code officer does not have to show his ID upon request.

      True   False

15.   It is okay to begin work on part of a plumbing system before the entire system has been approved.

      True   False

16.   The Code officer will notify the applicant, in writing, if denied a permit

      True   False

17.   Once a permit is issued it can be assigned to another person

      True   False

18. A set of plans must be kept at the job site.

True   False

19. All work performed must be done according to the plans, unless the homeowner feels it is necessary to alter them

True   False

20. All jurisdictions use the same plumbing codes

True   False

21. If a plumber is altering an existing system to add new plumbing, he or she must make all alterations in compliance with code requirements.

True   False

22. The code officer has the authority to alter the provisions of the plumbing code, so long as the homeowner agrees with it.

True   False

23. A variance is used in obtaining permission to alter permits that have already been approved.

True   False

24. A homeowner may apply for a supplementary permit, if he or she is having financial hardship.

True   False

25. Plumbing permits bear the signature of the code officer or authorized representative.

True   False

MULTIPLE-CHOICE ANSWERS

1. c	18. c	35. f	52. c	69. c
2. c	19. f	36. b	53. c	70. d
3. b	20. c	37. c	54. c	71. d
4. d	21. c	38. e	55. c	72. b
5. d	22. d	39. b	56. b	73. d
6. d	23. e	40. c	57. e	74. b
7. c	24. d	41. b	58. b	75. d
8. c	25. d	42. b	59. b	76. b
9. c	26. g	43. c	60. c	77. c
10. a	27. c	44. d	61. c	78. a
11. b	28. a	45. b	62. b	79. c
12. d	29. c	46. d	63. b	80. d
13. b	30. c	47. d	64. b	81. b
14. b	31. b	48. d	65. b	82. a
15. d	32. a	49. b	66. c	83. c
16. d	33. b	50. c	67. c	84. d
17. d	34. c	51. d	68. d	85. c

TRUE-FALSE ANSWERS

1. True	8. False	14. False	20. False
2. True	9. True	15. False	21. True
3. False	10. True	16. True	22. False
4. False	11. False	17. False	23. False
5. True	12. True	18. True	24. False
6. True	13. False	19. False	25. True
7. True			

Chapter 3

GENERAL REGULATIONS

In addition to many specific aspects of the plumbing code, there are a fair number of general code regulations you must know. These elements of the code are just as important as any other in a licensing exam. Not only are the general regulations required reading for the test; many of them are used frequently in the field. For example, the general regulations of the code dictate how the testing of a plumbing system must be done. If you are not already familiar with the general regulations, you will find that many will be used on a regular basis as you go about your plumbing work. Before we jump right into another mock exam, let's talk for a few moments about general regulations.

You might say that the general regulations are the foundation that the entire code is built around. Putting it in terms of construction, you could say that the general regulations are the footing for a structure. If you know anything about construction, you know how critical a good footing is to the construction of a building. Just as a footing is of paramount importance in construction, the general regulations of the plumbing code are equally important in the application of and compliance with the code.

What types of requirements do the general regulations cover? The list of subjects is not particularly long, but it is important. For example, the general regulations deal with health and safety. As a plumber, you might be thinking that this has to do with potable water and sewage. In reality, the general regulations go much deeper than that. Some of the regulations may not seem to be specific to plumbing, but they are related to work that is done in conjunction with the installation of plumbing. Let's talk about one such example.

When you read the general regulations, you will notice provisions that deal with cutting and notching structural members. This information would seem to apply more to carpenters than plumbers, but plumbers do a lot of cutting and notching. When a floor joist is blocking the path of a drain for a closet flange, something must be done. As the plumber on the job, you may have to decide what the right plan of action will be. Should you notch out the top of the joist, or should you cut the joist completely away and install headers on the severed ends? Suppose you are running a horizontal drain through floor joists and cannot keep the pipe in the center of the joists. Since the drain must be graded properly, you may find that your holes are getting close to the bottom of the joist. How close to the bottom of a joist can you drill a hole without causing structural or code problems? If you are in charge of plumbing the job, you'd better know. Drilling holes too close to the bottom of joists can result in some expensive and time-consuming repairs that you may be held responsible for. As you can see, there are parts of the general regulations that you may, at first, take for granted that are very important.

Let's look at another example. Assume that you are in charge of installing the building sewer. The sewer is required, due to the location of the main sewer, to run parallel to the footing of the structure. There is a backhoe operator on the job to dig the trench for you, but you must explain to the operator where and how the trench will be dug. Are you aware that there is a rule in the general regulations that pertains to this type of situation? Well, there is. The rule requires that the trench shall not extend below the line of a 45-degree angle to the load-bearing plane of the footing. The reason for this rule is fairly obvious: a more severe trench could weaken the structural integrity of the footing. Would you want to be responsible for digging a trench that undermined a foundation? Of course you wouldn't, and if you learn the general regulations, you can avoid such problems.

I've taken this brief detour from our sample tests to impress upon you the importance of the general regulations. During my career, I've seen a lot of plumbers and would-be plumbers, pay little attention to most of the general regulations. They assume the rules are not going to affect them. It's true that you may never use some of the rules found in the general regulations, but you never know which rule will come out of left field when you least expect it. You must learn the general regulations for your exam, and you should learn them to enjoy a safer and more productive time in the field. With that said, let's get started with the multiple-choice exam.

MULTIPLE-CHOICE EXAM

1.   It is common practice to make all measurements, such as the distance from one fixture drain to another, to what portion of a pipe

a.   The left side

b.   The right side

c.   The center

d.   The side closest to the adjoining fixture

2.   When a health or safety hazard exists on a premise due to an existing plumbing system or the lack of one, who is responsible to abate such a nuisance:

a.   A code officer

b.   A master plumber

c.   The property owner or an authorized agent

d.   A journeyman plumber

3.   A plumbing fixture that discharges only waste, such as a kitchen sink, can be drained into which of the following:

a.   A sanitary drainage system   b.   A French drain

c.   A gravel-lined pit   d.   Either A or B

4.   When altering a structure for the installation of plumbing, the job must be left in what condition:

a.   A safe and nonhazardous condition

b.   A condition equal to the way it was found

c.   A neat and clean condition

d.   A condition that is satisfactory to the property owner

5.   Trenching that runs parallel to the footing of a structure must not extend below the line of a _____ to the load-bearing plane of the footing. Fill in the blank:

a.   22 1/2-degree angle   b.   45-degree angle

c.   60-degree angle   d.   90-degree angle

> **! Code**alert
> A secondary drain or auxiliary drain pan shall be required for each cooling or evaporator coil or fuel-fired appliance that produces condensate, where damage to any building components will occur as a result of overflow from the equipment drain pan or stoppage in the condensate drain piping.

6.   Before waste products from manufacturing or industrial operations can be dumped into a public sewer, someone must determine if the waste will harm the sewer or sewage treatment facilities. Who has the authority to make this determination:

    a.   A master plumber

    b.   A code official

    c.   An authority authorized for such work

    d.   Either B or C

7.   Strainer plates on drain inlets shall be made with openings that are not larger than _____ in minumum dimension. Fill in the blank:

    a.   1/4 inch   b.   1/2 inch

    c.   3/4 inch   d.   1 inch

8.   What types of facilities must be provided for workers on construction sites?

    a.   Toilet facilities   b.   Rest facilities

    c.   Recreational facilities   d.   None of the above

9.   Most plumbing is prohibited from being installed in elevator shafts and elevator equipment rooms. There are, however, some exceptions; which of the following are an exception to the rule:

    a.   Floor drains   b.   Sumps and sump pumps

    c.   Both A and B   d.   None of the above

10.   Meter boxes are required to be constructed in such a way that unwanted wildlife or bugs cannot enter them. Which creature does the code specifically refer to:

    a.   Snakes   b.   Cats

    c.   Rodents   d.   Spiders

11.   When backfilling a trench, the fill material should be installed in layers. How thick should each layer be:

    a.   4 inches   b.   6 inches

    c.   8 inches   d.   12 inches

12.   Backfilling requires the use of loose earth that is free from which of the following:

    a.   Rocks   b.   Broken concrete

    c.   Frozen chunks   d.   All of the above

13.   When rock is encountered during trenching for a pipe, it must be removed to a level _____ below the grade line of the trench. Fill in the blank:

    a.   2 inches   b.   3 inches

    c.   4 inches   d.   6 inches

> **! Code**alert
> The auxiliary drain pan shall be equipped with a water-level detection device conforming to UL 508 that will shut off the equipment served prior to over-flow of the pan.

14. Water closets that are installed for public use and are not installed in a single-occupant toilet room equipped with a lockable door must occupy a separate compartment. These compartments must contain which of the following:

    a. A door
    b. Walls or partitions
    c. A handrail
    d. Both A and B

15. Piping installed as part of a plumbing system must be installed to prevent which of the following:

    a. Strains
    b. Stresses
    c. Both A and B
    d. None of the above

16. It may be necessary to protect plumbing pipes from which of the following:

    a. Expansion
    b. Contraction
    c. Structural settlement
    d. All of the above

17. Pipes that pass under or through a wall must be protected from:

    a. Rats
    b. Condensation
    c. Breakage
    d. All of the above

18. Any pipe that passes under a footing or through a foundation must be protected by a relieving arch or a _____. Fill in the blank:

    a. Sleeve
    b. Counterbalance
    c. Backwater valve
    d. Any of the above

19. When a sleeve is used for pipe protection, the sleeve must have a diameter at least _____ pipe sizes larger than the pipe being protected. Fill in the blank:

    a. 2
    b. 3
    c. Either A or B
    d. None of the above

20. There are times when pipes passing through concrete must be protected from corrosion; which of the following is a common form of protection in these circumstances?

    a. A protective coating
    b. A protective wrapping
    c. Both A and B
    d. None of the above

21. Which of the following types of pipe are not required to be protected from corrosion caused by passing through concrete?

    a. Copper   b. PVC

    c. ABS   d. Both B and C

22. If soft soil conditions are encountered when trenching for a pipe installation and solid ground is not readily available, it will be necessary to overexcavate the trench and backfill to the desired bed level to assure a solid base for bedding the pipe. When overexcavating is necessary, what is the maximum allowable depth of backfill layers prior to compaction:

    a. A minimum of one pipe size   b. A minimum of two pipe sizes

    c. 4 inches   d. 6 inches

23. Which of the following materials is suitable as stabilization backfill when a trench for piping has been overexcavated:

    a. Fine gravel   b. Crushed stone

    c. Concrete   d. All of the above

24. When a trench is backfilled, at what intervals must the fill material be compacted:

    a. Every 4 inches   b. Every 6 inches

    c. Every 8 inches   d. Every 12 inches

25. Which of the following types of pipe do not require the installation of nail plates to protect them:

    a. PVC   b. ABS

    c. Copper   d. Cast iron

26. Which of the following types of pipe do not require the installation of nail plates to protect them:

    a. PVC   b. PB

    c. Copper   d. Galvanized steel

27. If a copper pipe is installed within _____ from the edge of a stud, where it might be hit by a nail or a screw, the pipe must be protected with a nail plate. Choose the maximum depth that still requires protection:

    a. 3/4 inch   b. 1 inch

    c. 1 1/4 inch   d. 1 1/2 inch

28. What is the minimum thickness required for the construction of a nail plate:

    a. 0.035 inch   b. 0.045 inch

    c. 0.062 inch   d. 0.075 inch

29.   When a nail plate is installed to protect a pipe that is passing through a sole plate, the nail plate must extend how far above the sole plate:

    a.   1/2 inch

    b.   1 inch

    c.   1 1/2 inches

    d.   2 inches

30.   When a nail plate is installed to protect a pipe that is passing through a top plate, the nail plate must extend how far below the top plate:

    a.   1/2 inch

    b.   1 inch

    c.   1 1/2 inches

    d.   2 inches

31.   Pipes can be protected from freezing in unheated areas with the use of which of the following:

    a.   Insulation

    b.   Heat

    c.   Either A or B

    d.   None of the above

32.   At what intervals must buried pipe be supported:

    a.   4-foot intervals

    b.   6-foot intervals

    c.   10-foot intervals

    d.   For its entire length

33.   Pipes in trenches may not be covered with backfill until _____. Fill in the blank:

    a.   The pipe has been tested and approved by a code official

    b.   The backfill material is dry

    c.   The pipe has been tested for stress

    d.   All of the above

34.   When drilling holes in joists for the passage of pipes, the holes must not be drilled closer than the_____ allows. Fill in the blank:

    a.   The general contractor allows

    b.   The building code allows

    c.   The fire code allows

    d.   All of the above

! Codealert

A water level detection device conforming to UL508 shall be provided that will shut off the equipment served in the event that the primary drain is blocked. The device shall be installed in the primary drain line, the overflow drain line, or in the equipment-supplied drain pan located at a point higher than the primary drain line connection and below the overflow rim of such pan. Exception: Fuel fired appliances that automatically shut down operation in the event of a stoppage in the condensate drainage system.

35.   Notches on the ends of joists must not exceed the requirements of _____ . Fill in the blank:

a.   The property owner   b.   The general contractor

c.   The building code   d.   All of the above

36.   What must be done with the annular space between a pipe and a sleeve that protects the pipe:

a.   It must be left open

b.   It must be ventilated

c.   It must be protected from freezing

d.   It must be filled or tightly caulked

37.   The minimum depth for burying a water-service pipe is either 12 inches or _____ inches below the local frost line, whichever is deeper. Fill in the blank:

a.   4   b.   5

c.   6   d.   8

38.   Exterior openings made for plumbing must be made _____. Fill in the blank:

a.   Rodent-proof   b.   Watertight

c.   Both A and B   d.   None of the above

39.   Plumbing installations must not interfere with which of the following:

a.   Windows   b.   Doors

c.   Egress openings   d.   All of the above

40.   When water is used to test underground plumbing drains and vents, the use of a standpipe is required for the entry and observation of water used for the test. How tall is this standpipe required to be:

a.   4 feet   b.   5 feet

c.   8 feet   d.   10 feet

> **! Code**alert
>
> Water level monitoring devices: On downflow units, and all other coils that have no secondary drain and no means to install an auxiliary drain pan, a water level monitoring device shall be installed inside the primary drain pan. This device shall shut off the equipment served in the event that the primary drain becomes restricted. Externally installed devices and devices installed in the drain line shall not be permitted.

> ## ! Codealert
>
> Urinal partitions: Each urinal utilized by the public or employees shall occupy a separate area with walls or partitions to provide privacy. The construction of the wall or partition shall incorporate waterproof, smooth, readily cleanable, and nonabsorbent finish surfaces. The walls or partitions shall begin at a height not more than 12 inches (305 mm) from and extend not less than 60 inches (1524 mm) above the finished floor surface. The walls or partitions shall extend from the wall surface at each side of the urinal a minimum of 18 inches (457 mm) or to a point not less than 6 inches (152 mm) beyond the outermost front lip of the urinal measured from the finished back wall surface, whichever is greater.
>
> Exceptions: In a single occupancy or unisex toilet room with a lockable door, urinal partitions shall not be required. Toilet rooms located in day care and child care facilities, and containing two or more urinals, shall be permitted to have one urinal without partitions.

41. How long is water required to remain in a drainage system prior to an inspection of the joints:

    a. 10 minutes   b. 15 minutes

    c. 20 minutes   d. 30 minutes

42. If air is used to test a drain, waste, and vent system, what is the required test pressure:

    a. 5 pounds per square inch

    b. 10 pounds per square inch

    c. 15 pounds per square inch

    d. 18 pounds per square inch

43. For what period of time must a drain, waste, and vent system hold its test pressure to prove that there are no leaks in the system:

    a. 10 minutes   b. 15 minutes

    c. 20 minutes   d. 30 minutes

44. If water is used to test water-distribution piping, the water used in the test must be obtained from _____. Fill in the blank:

    a. A well   b. A sanitary drum

    c. A potable source   d. A licensed testing agency

45. The minimum depth for burying a sewer is _____. Fill in the blank:

    a. 6 inches   b. 12 inches

    c. 18 inches   d. 24 inches

You have completed the multiple-choice exam pertaining to general regulations. If you answered all of these questions correctly, you certainly have a good grasp of the general regulations applied in the plumbing code. Before you check your answers, however, take the fill in the blank and the true-false exam. After completing the true-false test you can check all of your answers.

FILL IN THE BLANK EXAM

1.   _____ are the rules or laws used to control an activity, such as plumbing.

2.   Every job that requires a permit also requires an _____.

3.   Alterations may include renovations, maintenance, _____and additions.

4.   Faucet replacements need to be made with _____materials.

5.   When a permit is required, it must be _____ before the work is started.

6.   The information required to submit plans will vary from _____to jurisdiction.

7.   When time is of the essence, it may be possible to obtain a _____ permit

8.   Code enforcement officers are frequently referred to as _____.

9.   Pipe damage can come in two forms,_____ and _____.

10.   Layers of backfill should not be more than _____ inches deep before they are compacted.

Cross off the answers as you use them

partial	obtained	inspectors	repairs	approved
immediate danger	jurisdiction	Regulations	inspection	
inspectors	long-term damage	six		

TRUE-FALSE EXAM

1. The plumbing code prohibits the threading of cast-iron pipe.
   True   False

2. Plumbing must be installed with due regard to the preservation of the strength of structural members.
   True   False

3. Side-inlet closet bends may be used to receive the discharge from a lavatory.
   True   False

4. Pipe trenches shall be filled in thin layers until there is a minimum of 12 inches of cover over the pipe.
   True   False

5. Inspection requests must be made at least 24 hours in advance of the time the inspection is desired.
   True   False

6. Water-distribution pipes must be tested with water at the working pressure of the system or with air at a pressure of 50 pounds per square inch.
   True   False

7. Condensate drains must be trapped as required by the equipment or appliance manufacturer.
   True   False

8. If an air test is conducted on a DWV system, the test pressure must be a minimum of 10 PSI.
   True   False

9. If a building is moved from one foundation to another, it is required that the entire plumbing system be tested in accordance with code requirements.
   True   False

10. Buried pipe must be supported at maximum intervals of four feet.
    Truc   Falsc

11. If plumbing installation is made in an area subject to flooding, special precautions must be taken.
    True   False

12. Flood-proofing must be in accordance with the requirements of the Town Office.
    True   False

13. Types of systems and equipment that may be allowed to be installed below a design flood elevation include water service pipes.

    True   False

14. When a pipe penetrates an exterior wall, it must pass through a sleeve.

    True   False

15. A sleeve must be at least 3 pipe sizes larger than the pipe.

    True   False

16. Soils are never capable of corroding pipes.

    True   False

17. ABS pipes need a maximum horizontal hanger spacing of at least 2 feet.

    True   False

18. Waste lines must not be less than .75 inch in internal diameter.

    True   False

19. Equipment efficiencies must conform to the International Energy Conduct Code.

    True   False

20. Many jobs require more than one inspection.

    True   False

21. When working with the plumbing code, you don't need to pay attention to the requirements as long as you know what you are doing.

    True   False

22. Regulations are only guidelines to follow if you have any questions.

    True   False

23. Generally, any existing condition that is not a hazard to health and safety is allowed to remain in an installation.

    True   False

24. You need to apply and be granted a permit before installing a faucet.

    True   False

25. Depending upon the nature of an alteration, you may need to apply for a permit.

    True   False

MULTIPLE-CHOICE ANSWERS

1.	c	10.	c	19.	a	28.	c	37.	c
2.	c	11.	b	20.	b	29.	d	38.	c
3.	a	12.	d	21.	d	30.	d	39.	d
4.	a	13.	b	22.	d	31.	c	40.	d
5.	b	14.	d	23.	a	32.	d	41.	b
6.	d	15.	c	24.	b	33.	a	42.	a
7.	b	16.	d	25.	d	34.	b	43.	b
8.	a	17.	c	26.	d	35.	c	44.	c
9.	b	18.	a	27.	d	36.	d	45.	b

FILL IN THE BLANK ANSWERS

1. Regulations
2. Inspection
3. Repairs
4. Approved
5. Obtained
6. Jurisdiction
7. Partial
8. Inspectors
9. Immediate Danger and long-term damage
10. Six inches

TRUE-FALSE ANSWERS

1.	True	8.	False	14.	True	20.	True
2.	True	9.	True	15.	False	21.	False
3.	False	10.	False	16.	False	22.	False
4.	True	11.	True	17.	False	23.	True
5.	True	12.	False	18.	True	24.	False
6.	True	13.	True	19.	False	25.	True
7.	True						

TABLE 3.1 Products and materials requiring third-party testing and third-party certification. *Copyright 2006, International Code Council, Inc., Falls Church, Virginia. Reproduced with permission. All rights reserved.*

PRODUCT OR MATERIAL	THIRD-PARTY CERTIFIED	THIRD-PARTY TESTED
Potable water supply system components and potable water fixture fittings	Required	—
Sanitary drainage and vent system components	Plastic pipe, fittings and pipe-related components	All others
Waste fixture fittings	Plastic pipe, fittings and pipe-related components	All others
Storm drainage system components	Plastic pipe, fittings and pipe-related components	All others
Plumbing fixtures	—	Required
Plumbing appliances	Required	—
Backflow prevention devices	Required	—
Water distribution system safety devices	Required	—
Special waste system components	—	Required
Subsoil drainage system components	—	Required

TABLE 3.2 Hanger spacing. *Copyright 2006, International Code Council, Inc., Falls Church, Virginia. Reproduced with permission. All rights reserved.*

PIPING MATERIAL	MAXIMUM HORIZONTAL SPACING (feet)	MAXIMUM VERTICAL SPACING (feet)
ABS pipe	4	10[b]
Aluminum tubing	10	15
Brass pipe	10	10
Cast-iron pipe	5[a]	15
Copper or copper-alloy pipe	12	10
Copper or copper-alloy tubing, 1$\frac{1}{4}$-inch diameter and smaller	6	10
Copper or copper-alloy tubing, 1$\frac{1}{2}$-inch diameter and larger	10	10
Cross-linked polyethylene (PEX) pipe	2.67 (32 inches)	10[b]
Cross-linked polyethylene/ aluminum/cross-linked polyethylene (PEX-AL-PEX) pipe	2.67 (32 inches)	4[b]
CPVC pipe or tubing, 1 inch and smaller	3	10[b]
CPVC pipe or tubing, 1$\frac{1}{4}$ inches and larger	4	10[b]
Steel pipe	12	15
Lead pipe	Continuous	4
PB pipe or tubing	2.67 (32 inches)	4
Polyethylene/aluminum/ polyethylene (PE-AL-PE) pipe	2.67 (32 inches)	4[b]
Polypropylene (PP) pipe or tubing 1 inch and smaller	2.67 (32 inches)	10[b]
Polypropylene (PP) pipe or tubing, 1$\frac{1}{4}$ inches and larger	4	10[b]
PVC pipe	4	10[b]
Stainless steel drainage systems	10	10[b]

For SI: 1 inch = 25.4 mm, 1 foot = 304.8 mm.

a. The maximum horizontal spacing of cast-iron pipe hangers shall be increased to 10 feet where 10-foot lengths of pipe are installed.

b. Midstory guide for sizes 2 inches and smaller.

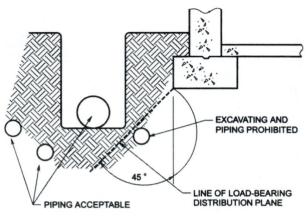

For SI: 1 degree = 0.01745 rad.

FIGURE 3.1   Protecting a footing during excavation and pipe installation. *Copyright 2002, International Code Council, Inc., Falls Church, Virginia. Reproduced with permission. All rights reserved.*

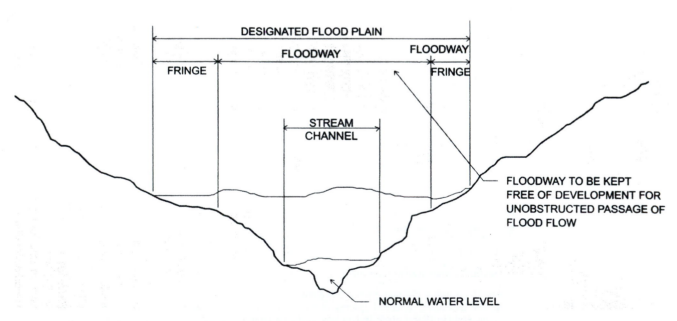

FIGURE 3.2   Cross section of a flood plain. *Copyright 2002, International Code Council, Inc., Falls Church, Virginia. Reproduced with permission. All rights reserved.*

Chapter 4
FIXTURES

The job of choosing and installing proper plumbing fixtures is much more complicated than many people realize. A lot of plumbers think that fixture choice is up to the property owners for whom the work is being done. To some extent, this is correct, but the subject involves more than meets the average eye. Master plumbers and architects are typically responsible for determining the minimum fixture requirements for a building. In residential plumbing, the required fixtures are not difficult to list. However, in commercial plumbing, knowing what types of fixtures are required and in what numbers gets more difficult. To expand on this, let me give you a few examples of what a master plumber must be able to determine.

Let's assume that you are asked to design a plumbing system for a factory. Would you know that you must have one water closet for each 100 people? Did you know that one lavatory would be required for every 100 people in the building? There are no bathtubs required, but there must be at least one drinking fountain for every 400 people. There is also a requirement that that one service sink be installed. Now, let's compare this fixture selection to that of a building used as a dormitory.

Dormitories must be equipped with one water closet for every 10 people using the building. The same ratio applies to lavatories. There must be one bathtub or shower for every 8 people. A drinking fountain must be installed for every 100 people. And one service sink be installed in the building. Since you are preparing for a plumber's exam, you will have to learn this type of information. For now, we will begin your introduction to plumbing fixtures with a multiple-choice exam.

MULTIPLE-CHOICE EXAM

1. A device that is considered to be a special class of plumbing fixtures and is intended to perform a special function is :

    a. A bidet   b. A combination fixture

    c. A plumbing appliance d. None of the above

2. Fixtures that depend upon one or more energized components, such as motors, controls, heating elements, or pressure or temperature sensing elements, are usually called _____. Fill in the blank:

    a. Plumbing appliances   b. Mechanized fixtures

    c. Complex fixtures   d. None of the above

3. A _____ doesn't require an additional water supply or add discharge load to a fixture or drainage system. Fill in the blank:

    a. Plumbing appliance   b. Plumbing appurtenance

    c. Plumbing fixture   d. None of the above

4. A receptacle or device that is either permanently or temporarily connected to the water-distribution system of the premises and takes water from it is called _____. Fill in the blank:

    a. A plumbing appurtenance b. A plumbing fixture

    c. Both A and B   d. Neither A or B

5. A receptacle or device that is either permanently or temporarily connected to a plumbing system that discharges waste water, liquid-borne waste materials, or sewage either directly or indirectly to the drainage system is called _____. Fill in the blank:

    a. A plumbing appurtenance
    b. A plumbing fixture
    c. Both A and B
    d. Neither A or B

6. A receptacle or device that requires both a water-supply connection and a discharge to a drainage system is called a:

    a. A plumbing appurtenance
    b. A plumbing fixture
    c. Both A and B
    d. A plumbing conservation device

7. A water closet that is installed in such a way that it does not touch a floor is called:

    a. A carrier water closet
    b. A hanger water closet
    c. A wall-hung water closet
    d. None of the above

8. A bathtub that is equipped and fitted with a circulation piping system, pump, and similar appurtenances and is so designed to accept, circulate, and discharge bathtub water is called a:

    a. Spa
    b. Whirlpool bathtub
    c. Either A or B
    d. Both A and B

9. Which of the following locations would be considered to have fixtures meant for public use:

    a. Toilet rooms in schools
    b. Toilet rooms in gymnasiums
    c. Both A and B
    d. Neither A or B

10. Which of the following locations would be considered to have fixtures meant for public use:

    a. Toilet rooms in hotels
    b. Toilet rooms in apartments
    c. Both A and B
    d. Neither A or B

11. Which of the following locations would be considered to have fixtures meant for public use:

    a. Toilet rooms in railroad stations
    b. Toilet rooms in bars
    c. Both A and B
    d. Neither A or B

12. Which of the following locations would be considered to have fixtures meant for public use:

    a. Toilet rooms in public buildings
    b. Toilet rooms in residences
    c. Both A and B
    d. Neither A or B

13.   Which of the following locations would be considered to have fixtures meant for public use:

    a.   Toilet rooms in public comfort stations

    b.   Toilet rooms in apartments

    c.   Both A and B

    d.   Neither A or B

14.   Which of the following locations would be considered to have fixtures meant for private use:

    a. Toilet rooms in public comfort stations

    b. Toilet rooms in apartments

    c. Both A and B

    d. Neither A or B

15.   Which of the following locations would be considered to have fixtures meant for private use:

    a.   Toilet rooms in residences

    b.   Toilet rooms in apartments

    c.   Both A and B

    d.   Neither A or B

16.   All plumbing fixtures and drains that receive or discharge sewage are required to be connected to:

    a.   A storm drain   b.   A gray-water drain

    c.   A sanitary drainage system   d.   None of the above

17.   Separate facilities are often required for men and women using a building. Which of the following types of buildings are exempt from this rule:

    a.   Residential properties

    b.   Business properties where 15 people or fewer are employed

    c.   Buildings with a total occupant load, including both employees and customers, of 15 or fewer in which food or beverages are served for consumption within the structure.

    d.   All of the above

18.   Under normal conditions, the ratio of fixtures provided for females and males should be:

    a.   Evenly distributed with 50 percent for each sex

    b.   Distributed with 75 percent of the fixtures allotted for females

    c.   Distributed with 75 percent of the fixtures allotted for males

    d.   None of the above

19. In most cases, access to toilet facilities for employees must be:

    a. Within 50 feet of the employees' work space

    b. Accessible from the employees' regular working area

    c. Both A and B

    d. Neither A or B

20. Toilet facilities for employees, in most cases, must be located not more than _____ above or below the employees' regular work area. Fill in the blank:

    a. 1 story

    b. 2 stories

    c. 18 feet

    d. None of the above

21. Toilet facilities for employees, in most cases, must be located in such a way that employees will not have to travel more than _____ feet to reach them. Fill in the blank:

    a. 50

    b. 100

    c. 200

    d. 500

22. Some types of buildings require toilet facilities for employees. The facilities may be separate or the same as are used for public customers. Which of the following uses require such facilities:

    a. Restaurants

    b. Nightclubs

    c. Both A and B

    d. None of the above

23. Some types of buildings require toilet facilities for employees. The facilities may be separate or the same as are used for public customers. Which of the following uses require such facilities:

    a. Places of public assembly

    b. Mercantile occupancies

    c. Both A and B

    d. None of the above

24. Customers, patrons, and visitors to certain types of buildings are entitled to toilet facilities. Which of the following types of building uses will trigger the need for such toilet facilities:

    a. Restaurants

    b. Mercantile occupancies

    c. Both A and B

    d. None of the above

25. Customers, patrons, and visitors to certain types of buildings are entitled to toilet facilities. Which of the following types of building uses will trigger the need for such toilet facilities?

    a. Restaurants

    b. Nightclubs

    c. Both A and B

    d. None of the above

26. Customers, patrons, and visitors to certain types of buildings are entitled to toilet facilities. Which of the following types of building uses will trigger the need for such toilet facilities:

    a. Restaurants
    b. Places of assembly
    c. Both A and B
    d. None of the above

27. Toilet rooms and bathing facilities containing fixtures for occupants of a building that is required to be accessible to physically disadvantaged persons must have at least _____ of each type. Fill in the blank:

    a. 1
    b. 2
    c. 3
    d. 4

28. Toilet rooms and bathing facilities containing fixtures for occupants of a building that is required to be accessible to physically disadvantaged persons must have at least _____ wheelchair-accessible compartment for a water closet. Fill in the blank:

    a. 1
    b. 2
    c. 3
    d. 4

29. When 6 or more water-closet compartments are provided in a toilet room or bathing facility, at least _____ ambulatory-accessible compartment must be provided in addition to the required wheelchair-accessible compartment. Fill in the blank:

    a. 1
    b. 2
    c. 3
    d. 4

30. Where drinking fountains are installed on each floor of buildings required to be accessible to physically disadvantaged persons, _____ percent, but not less than one drinking fountain or water cooler per floor, shall be accessible to and usable by physically disadvantaged persons. Fill in the blank:

    a. 25
    b. 33 1/3
    c. 50
    d. 75

31. When showers are installed in gymnasiums in conjunction with sports activities, in buildings required to be adapted to meet the needs of physically disadvantaged persons, a minimum of _____ shall be accessible to and usable by physically disabled persons:

    a. 1 shower
    b. 2 showers
    c. 3 showers
    d. 4 showers

32. The minimum net clear opening to a water-closet compartment shall be:

    a. 30 inches
    b. 32 inches
    c. 36 inches
    d. 42 inches

33. Doors installed on water-closet compartments must swing _____. Fill in the blank:

    a. In, towards the toilet

    b. Out, away from the toilet

    c. Either A or B

    d. Neither A or B

34. The minimum clear floor space in front of a door to a water-closet compartment shall be:

    a. 32 x 30 inches   b. 30 x 36 inches

    c. 36 x 42 inches   d. 30 x 48 inches

35. What is the minimum distance of clear space required from the center of a water closet to the closest wall:

    a. 12 inches   b. 15 inches

    c. 18 inches   d. 30 inches

36. What is the minimum distance of clear space required from the center of a lavatory to the closest wall:

    a. 12 inches   b. 15 inches

    c. 18 inches   d. 30 inches

37. What is the minimum distance of clear space required from the center of a bidet to the closest wall:

    a. 12 inches   b. 15 inches

    c. 18 inches   d. 30 inches

38. How much clear space is required in front of a water closet:

    a. 12 inches   b. 15 inches

    c. 21 inches   d. 30 inches

39. How much clear space is required in front of a lavatory:

    a. 12 inches   b. 15 inches

    c. 16 inches   d. 21 inches

40. How much clear space is required in front of a bidet:

    a. 12 inches   b. 15 inches

    c. 21 inches   d. 30 inches

41. When water closets are installed side by side, what is the minimum distance required between the center-lines of the adjacent fixtures:

    a. 12 inches   b. 15 inches

    c. 18 inches   d. 30 inches

42.   When lavatories are installed side by side, what is the minimum distance required between the centerlines of the adjacent fixtures:

a.   12 inches   b.   15 inches

c.   18 inches   d.   30 inches

43.   When bidets are installed side by side, what is the minimum distance required between the centerlines of the adjacent fixtures:

a.   12 inches   b.   15 inches

c.   18 inches   d.   30 inches

44.   What is the minimum distance of clear space required from the center of a urinal to the closest wall:

a.   12 inches   b.   15 inches

c.   18 inches   d.   30 inches

45.   Fixtures equipped with concealed slip-joint connections, as in the case of a tub waste and overflow, must be provided with _____. Fill in the blank:

a.   An access panel   b.   A utility space

c.   Either A or B   d.   None of the above

46.   Fixtures equipped with concealed solvent-cemented joint connections, as in the case of a tub waste and overflow, must be provided with _____. Fill in the blank:

a.   An access panel   b.   A utility space

c.   Either A or B   d.   None of the above

47.   Water supplies to automatic clothes washers must be protected against backflow by a backflow preventer or _____. Fill in the blank:

a.   An air break   b.   An air gap

c.   Either A or B   d.   Neither A or B

48.   The waste discharge from an automatic washer must be through _____. Fill in the blank:

a.   An air break   b.   An air gap

c.   Either A or B   d.   Neither A or B

49.   The minimum size allowed for waste and overflow for a bathtub is:

a.   1 1/4 inches   b.   1 1/2 inches

c.   2 inches   d.   2 1/2 inches

50. Water supplies to bidets must be protected against backflow by a backflow preventer or _____. Fill in the blank:

   a. An air break   b. An air gap

   c. Either A or B   d. Neither A or B

51. Water supplies to dishwashers must be protected against backflow by a backflow preventer or _____. Fill in the blank:

   a. An air break   b. An air gap

   c. Either A or B   d. Neither A or B

52. Which of the following is a suitable place to accept the discharge from a dishwasher:

   a. A trap

   b. A wye-branch tailpiece of a trapped kitchen sink fixture

   c. A food-waste grinder

   d. All of the above

53. Which of the following is a suitable place to accept the discharge from a dishwasher:

   a. A trapped standpipe

   b. A food-waste grinder

   c. A tailpiece of a kitchen sink

   d. All of the above

54. Drinking fountains may not be installed in which of the following locations:

   a. Private bathrooms   b. Public bathrooms

   c. Both A and B   d. Neither A or B

55. Emergency showers must be provided with a supply of _____. Fill in the blank:

   a. Cold water   b. Hot water

   c. Both A and B   d. None of the above

56. Which of the following is not required for an emergency shower:

   a. A hot-water supply   b. A waste connection

   c. Both A and B   d. None of the above

57. Floor drains must have a minimum size of:

   a. 1 1/2 inches   b. 2 inches

   c. 3 inches   d. 4 inches

58. The strainers that cover the traps of floor drains must be _____. Fill in the blank:

a. Constructed of brass   b. Constructed of cast iron

c. Removable   d. Both A and C

59. The minimum size for a drain that will accept the waste from a domestic food-waste grinder is:

a. 1 1/4 inches   b. 1 1/2 inches

c. 2 inches   d. 3 inches

60. The minimum size for a drain that will accept the waste from a commercial food-waste grinder is:

a. 1 1/4 inches   b. 1 1/2 inches

c. 2 inches   d. 3 inches

61. Commercial dishwashers must _____. Fill in the blank:

a. Discharge through the tailpiece of a sink

b. Be trapped and connected to the drainage system separately from any other fixture or sink compartments

c. Discharge, by way of an air gap, into a deep sink bowl

d. None of the above

62. Water supplies to garbage-can washers must be protected against backflow by a backflow preventer or _____. Fill in the blank:

a. An air break   b. An air gap

c. Either A or B   d. Neither A or B

63. Garbage-can washers must _____. Fill in the blank:

a. Discharge through a sand interceptor

b. Be trapped separately

c. Both A and B

d. Either A or B

64. Receptors that receive waste from garbage-can washers must _____. Fill in the blank:

a. Discharge through a sand interceptor

b. Be equipped with removable baskets or strainers

c. Both A and B

d. Either A or B

65. The minimum size of a waste outlet for a laundry tray is:

a. 1 1/4 inches

b. 1 1/2 inches

c. 2 inches

d. 3 inches

66. Each compartment of a laundry tray must be equipped with _____. Fill in the blank:

a. A crossbar to restrict the clear opening of the waste outlet

b. A strainer to restrict the clear opening of the waste outlet

c. Both A and B

d. Either A or B

67. What effect does a food-waste grinder have on the fixture-unit load of a sanitary drainage system:

a. It increases it

b. It increases it by three fixture units

c. It has no effect on the fixture-unit load

d. Both A and B

68. The minimum size of a waste outlet for a lavatory is:

a. 1 1/4 inches

b. 1 1/2 inches

c. 2 inches

d. 3 inches

69. Which of the following is suitable for restricting the clear opening of a waste outlet in a lavatory:

a. A strainer

b. A pop-up plug

c. A crossbar

d. Any of the above

70. The minimum size for a waste outlet that serves a shower and is not part of a tub-shower combination is:

a. 1 1/4 inches

b. 1 1/2 inches

c. 2 inches

d. 3 inches

71. What is the minimum diameter size of a shower strainer?

a. 1 1/2 inches

b. 2 inches

c. 3 inches

d. 4 inches

72. What is the minimum size of the openings required in strainers for shower drains?

a. 1/4 inch

b. 1/2 inch

c. 3/4 inch

d. 1 inch

73. Shower compartments must contain a minimum of _____ square inches of interior cross-sectional area. Fill in the blank:

a. 600   b. 900

c. 1,100   d. 1,500

74. What is the minimum dimension of a shower when measured from the finished interior of the compartment, exclusive of the fixture valves, shower heads, soap dishes, and safety grab bars:

a. 30 inches   b. 32 inches

c. 34 inches   d. 36 inches

75. When shower pans are formed from pan material, the sides of such pans must turn up on all sides by at least _____ inches. Fill in the blank:

a. 2   b. 2 1/2

c. 3   d. 4

76. The minimum size of a waste outlet for a sink is:

a. 1 1/4 inches   b. 1 1/2 inches

c. 2 inches   d. 3 inches

77. When sinks are equipped with food-waste grinders, the sink's waste opening must have a minimum diameter of _____ inches. Fill in the blank:

a. 1 1/2   b. 2

c. 3   d. 3 1/2

78. When calculating the required fixtures for a specific building, it is often acceptable to substitute urinals for some water closets. However, in no case can urinals be substituted for more than _____ percent of the required water closets. Fill in the blank:

a. 25   b. 40

c. 55   d. 67

79. Water closets installed for use by the public must be of a(n) _____ type. Fill in the blank:

a. Flush-tank   b. Round-front

c. Elongated   d. None of the above

80. Water closets installed for use by the public or by employees must have _____. Fill in the blank:

a. Black seats   b. Hinged round-front seats

c. Hinged closed-front seats   d. Hinged open-front seats

81. Water closets that depend on trap siphonage to discharge the fixture contents to the drainage system must be provided with _____. Fill in the blank:

    a. A flushometer valve
    b. A flush tank
    c. Either A or B
    d. Both A and B

82. Urinals that depend on trap siphonage to discharge the fixture contents to the drainage system must be provided with _____. Fill in the blank:

    a. A flushometer valve
    b. A flush tank
    c. Either A or B
    d. Both A and B

83. Clinical sinks that depend on trap siphonage to discharge the fixture contents to the drainage system must be provided with _____. Fill in the blank:

    a. A flushometer valve
    b. A flush tank
    c. Either A or B
    d. Neither A or B

84. Flushing devices are not allowed to serve more than_____. Fill in the blank:

    a. 1 fixture
    b. 2 fixtures
    c. 2 fixtures of the same type
    d. None of the above

85. When installing a whirlpool tub, a plumber must be sure that the pump for the whirlpool is _____ the weir of the fixture trap. Fill in the blank:

    a. Below
    b. Above
    c. On the same level as
    d. None of the above

We are finally finished with the multiple-choice section of this chapter. I apologize for the length of this exam, but the subject of plumbing fixtures is a complicated one. It is also a very important part of your licensing exam. As you can tell by the volume of material covered in this exam, you will have a lot to study. If you're still with me at this point, you deserve a break before we move into the fill in the blank exam. Take a break, rest your eyes, walk around, and then proceed to the next section.

FILL IN THE BLANK EXAM

1.   A nightclub must have one toilet for every _____ people.

2.   Drinking fountains are not required in _____.

3.   There are some special _____ pertaining to employee and customer facilities.

4.   The distance an employee is required to walk to the facilities may not exceed _____ feet.

5.   The accessible route to public facilities shall not pass through a _____.

6.   The door to a privacy stall for a handicap toilet must provide a minimum of _____ inches of clear space for wheelchair access.

7.   It is mandatory that the door to a privacy handicap stall open _____.

8.   When a drainage connection is made with removable connections, the connections must be _____.

9.   Laundry trays are required to have 1.5 inch drains that should be equipped with _____.

10.   You may not install a water faucet in a room that contains a water _____.

Cross off the answers as you use them

32	requirements	40	accessible	closet
kitchen	500	restaurants	outward	crossbars

TRUE-FALSE EXAM

1. The bowls of water closets installed for public use must be elongated.

True   False

2. An access panel must be provided wherever concealed slip-joint connections exist.

True   False

3. The minimum size of a drain that serves a bathtub is 2 inches.

True   False

4. Backflow protection is not required for ornamental ponds.

True   False

5. When urinals are installed side by side, there should be a minimum of 30 inches from the center of one urinal to the center of next closest urinal.

True   False

6. Floor drains are not considered to be plumbing fixtures.

True   False

7. All shower compartments must have a minimum finished interior space of 1024 square inches.

True   False

8. All whirlpool tubs must be equipped with a removable panel.

True   False

9. The circulation pumps used with whirlpool tubs must be installed above the crown weir of the trap.

True   False

10. The minimum size for a lavatory drain is 1 1/2 inches.

True   False

11. Backflow protection must be provided for all plumbing fixtures.

True   False

12. The minimum size for a sink drain is 1 1/4 inches.

True   False

13. A drinking fountain must be installed in all public toilet rooms that are accessible to individuals with physical restrictions.

    True   False

14. The strainers installed on floor drains must be removable.

    True   False

15. The minimum drain size for a commercial food-waste grinder is 3 inches.

    True   False

16. The minimum drain size for a floor drain is 2 inches.

    True   False

17. All dishwashers must be connected directly to a sanitary drainage system.

    True   False

18. Domestic food-waste grinders must be equipped with a hot-water supply.

    True   False

19. Dormitories are required to have a minimum of one drinking fountain for each 100 people using the facility.

    True   False

20. All single-family residential dwellings are required to have at least one hookup for an automatic clothes-washing machine.

    True   False

21. Buildings used as day nurseries are required to have a minimum of one water closet for every 15 people using the facility.

    True   False

22. Up to three garbage-can washers may discharge into a common trap so long as the trap is not more than 30 inches from a vent.

    True   False

23. Drinking fountains may not be installed in private bathrooms.

    True   False

24. The waste connection for an automatic clothes washer must enter the drainage system through an air gap.

    True   False

25.   A 2-inch drain is required when a domestic food-waste grinder is installed.
     True   False

26.   The amount of water used by a urinal, in a single flush, should be limited to a maximum of 1.5 gallons.
     True   False

27.   Hospitals are required to have at least three water services.
     True   False

28.   Water coolers and fountains are a must have for toilet facilities.
     True   False

29.   Sheet copper cannot be used as a shower pan liner.
     True   False

30.   Hot and cold water can be piped to either left or right.
     True   False

31.   Valves or faucets used for showers must be designed to provide protection from scalding.
     True   False

32.   When installing handicap plumbing, you must combine the plumbing code with the building code.
     True   False

33.   The minimum number and type of fixtures for a single-family dwelling is two or more.
     True   False

34.   Most buildings frequented by the public are required to have handicap accessible fixtures.
     True   False

35.   Flush-valves are not required to be equipped with accessible vacuum breakers.
     True   False

36.   The minimum number of washing machine hookups for a single-family is one unit.
     True   False

37.   The minimum requirements for a multi-family and a single-family dwelling are the same except for the laundry hookup.
     True   False

38.   All types of buildings require separate facilities.
     True   False

39.   Single family and most residential multi-family dwellings are exempt from handicap requirements.
True   False

40.   Handicap fixtures are specially designed for people with less physical ability than the general public.
True   False

You have finished your test on the subject of plumbing fixtures. Check your answers with the correct answers below. Allow ample time to go over the questions and answers carefully. As you can tell by the length of this chapter, plumbing fixtures could amount to a large percentage of your licensing exam. When you are comfortable with your knowledge of fixtures, move onto the next chapter.

MULTIPLE-CHOICE ANSWERS

1. d	18. a	35. b	52. d	69. d
2. a	19. b	36. b	53. d	70. c
3. b	20. a	37. b	54. c	71. b
4. d	21. d	38. c	55. a	72. a
5. b	22. c	39. d	56. c	73. b
6. b	23. a	40. c	57. b	74. a
7. c	24. c	41. d	58. c	75. a
8. b	25. c	42. d	59. b	76. b
9. c	26. c	43. d	60. c	77. a
10. a	27. a	44. b	61. b	78. d
11. c	28. a	45. a	62. b	79. c
12. a	29. a	46. d	63. b	80. d
13. a	30. c	47. b	64. b	81. a
14. b	31. a	48. b	65. b	82. a
15. c	32. a	49. b	66. d	83. a
16. c	33. a	50. b	67. c	84. a
17. d	34. d	51. b	68. a	85. b

FILL IN THE BLANK ANSWERS

1. 40
2. restaurants
3. requirements
4. 500
5. kitchen
6. 32
7. outward
8. accessible
9. crossbars
10. closet

TRUE-FALSE ANSWERS

1.	True	11.	False	21.	True	31.	True
2.	True	12.	False	22.	False	32.	True
3.	False	13.	False	23.	True	33.	False
4.	False	14.	True	24.	True	34.	True
5.	True	15.	False	25.	False	35.	False
6.	False	16.	True	26.	True	36.	True
7.	False	17.	False	27.	False	37.	True
8.	True	18.	False	28.	False	38.	False
9.	False	19.	True	29.	False	39.	True
10.	False	20.	True	30.	False	40.	True

TABLE 4.1 Minimum number of required plumbing fixtures.a Copyright 2006, International Code Council, Inc., Falls Church, Virginia. Reproduced with permission. All rights reserved.

| NO. | CLASSIFICATION | OCCUPANCY | DESCRIPTION | WATER CLOSETS (URINALS SEE SECTION 419.2) | | LAVATORIES | | BATHTUBS/ SHOWERS | DRINKING FOUNTAIN (SEE SECTION 410.1) | OTHER |
				MALE	FEMALE	MALE	FEMALE			
1	Assembly	A-1d	Theaters and other buildings for the performing arts and motion pictures	1 per 125	1 per 65	1 per 200	1 per 200	—	1 per 500	1 service sink
		A-2d	Nightclubs, bars, taverns, dance halls and buildings for similar purposes	1 per 40	1 per 40	1 per 75	1 per 75	—	1 per 500	1 service sink
			Restaurants, banquet halls and food courts	1 per 75	1 per 75	1 per 200	1 per 200	—	1 per 500	1 service sink
		A-3d	Auditoriums without permanent seating, art galleries, exhibition halls, museums, lecture halls, libraries, arcades and gymnasiums	1 per 125	1 per 65	1 per 200	1 per 200	—	1 per 500	1 service sink
			Passenger terminals and transportation facilities	1 per 500	1 per 500	1 per 750	1 per 750	—	1 per 1,000	1 service sink
			Places of worship and other religious services.	1 per 150	1 per 75	1 per 200	1 per 200	—	1 per 1,000	1 service sink

NO.	CLASSIFICATION	OCCUPANCY	DESCRIPTION	WATER CLOSETS		LAVATORIES		BATHTUBS/ SHOWERS	DRINKING FOUNTAIN	OTHER
				MALE	FEMALE	MALE	FEMALE			
		A-4	Coliseums, arenas, skating rinks, pools and tennis courts for indoor sporting events and activities	1 per 75 for the first 1,500 and 1 per 120 for the remainder exceeding 1,500	1 per 40 for the first 1,500 and 1 per 60 for the remainder exceeding 1,500	1 per 200	1 per 150	—	1 per 1,000	1 service sink
		A-5	Stadiums, amusement parks, bleachers and grandstands for outdoor sporting events and activities	1 per 75 for the first 1,500 and 1 per 120 for the remainder exceeding 1,500	1 per 40 for the first 1,500 and 1 per 60 for the remainder exceeding 1,500	1 per 200	1 per 150	—	1 per 1,000	1 service sink
2	Business	B	Buildings for the transaction of business, professional services, other services involving merchandise, office buildings, banks, light industrial and similar uses	1 per 25 for the first 50 and 1 per 50 for the remainder exceeding 50		1 per 40 for the first 80 and 1 per 80 for the remainder exceeding 80		—	1 per 100	1 service sink

(continued)

TABLE 4.1 (continued)   Minimum number of required plumbing fixtures[a]. *Copyright 2006, International Code Council, Inc., Falls Church, Virginia. Reproduced with permission. All rights reserved.*

No.	Classification	Occupancy	Description					
3	Educational	E	Educational facilities	1 per 50	1 per 50	—	1 per 100	1 service sink
4	Factory and industrial	F-1 and F-2	Structures in which occupants are engaged in work fabricating, assembly or processing of products or materials	1 per 100	1 per 100	(see Section 411)	1 per 400	1 service sink
5	Institutional	I-1	Residential care	1 per 10	1 per 10	1 per 8	1 per 100	1 service sink
		I-2	Hospitals, ambulatory nursing home patients[b]	1 per room[c]	1 per room[c]	1 per 15	1 per 100	1 service sink per floor
			Employees, other than residential care[b]	1 per 25	1 per 35	—	1 per 100	—
			Visitors, other than residential care	1 per 75	1 per 100	—	1 per 500	—
		I-3	Prisons[b]	1 per cell	1 per cell	1 per 15	1 per 100	1 service sink
		I-3	Reformatories, detention centers, and correctional centers[b]	1 per 15	1 per 15	1 per 15	1 per 100	1 service sink
		I-4	Adult day care and child care	1 per 15	—	—	1 per 100	1 service sink

NO.	CLASSIFICATION	OCCUPANCY	DESCRIPTION	WATER CLOSETS		LAVATORIES		BATHTUBS/SHOWERS	DRINKING FOUNTAIN	OTHER
				MALE	FEMALE	MALE	FEMALE			
6	Mercantile	M	Retail stores, service stations, shops, salesrooms, markets and shopping centers	1 per 500		1 per 750		—	1 per 1,000	1 service sink
7	Residential	R-1	Hotels, motels, boarding houses (transient)	1 per sleeping unit		1 per sleeping unit		1 per sleeping unit	—	1 service sink
		R-2	Dormitories, fraternities, sororities and boarding houses (not transient)	1 per 10		1 per 10		1 per 8	1 per 100	1 service sink
		R-2	Apartment house	1 per dwelling unit		1 per dwelling unit		1 per dwelling unit	—	1 kitchen sink per dwelling unit; 1 automatic clothes washer connection per 20 dwelling units

(continued)

TABLE 4.1 (continued)   Minimum number of required plumbing fixtures[a]. *Copyright 2006, International Code Council, Inc., Falls Church, Virginia. Reproduced with permission. All rights reserved.*

7	Residential *(continued)*	R-3	One- and two-family dwellings	1 per dwelling unit	1 per dwelling unit	1 per dwelling unit	—	1 kitchen sink per dwelling unit; 1 automatic clothes washer connection per dwelling unit
		R-4	Residential care/assisted living facilities	1 per 10	1 per 10	1 per 8	1 per 100	1 service sink
8	Storage	S-1 S-2	Structures for the storage of goods, warehouses, storehouse and freight depots. Low and Moderate Hazard.	1 per 100	1 per 100	See Section 411	1 per 1,000	1 service sink

a. The fixtures shown are based on one fixture being the minimum required for the number of persons indicated or any fraction of the number of persons indicated. The number of occupants shall be determined by the *International Building Code.*

b. Toilet facilities for employees shall be separate from facilities for inmates or patients.

c. A single-occupant toilet room with one water closet and one lavatory serving not more than two adjacent patient sleeping units shall be permitted where such room is provided with direct access from each patient room and with provisions for privacy.

d. The occupant load for seasonal outdoor seating and entertainment areas shall be included when determining the minimum number of facilities required.

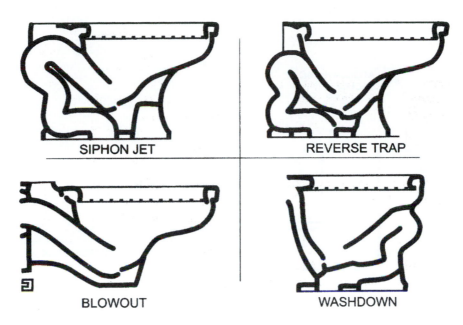

FIGURE 4.1 Types of closet bowl designs. *Copyright 2002, International Code Council, Inc., Falls Church, Virginia. Reproduced with permission. All rights reserved.*

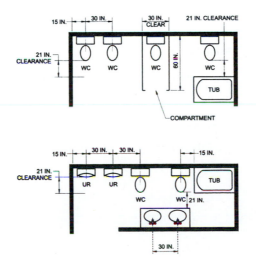

FIGURE 4.2 Fixture clearance. *Copyright 2006, International Code Council, Inc., Falls Church, Virginia. Reproduced with permission. All rights reserved.*

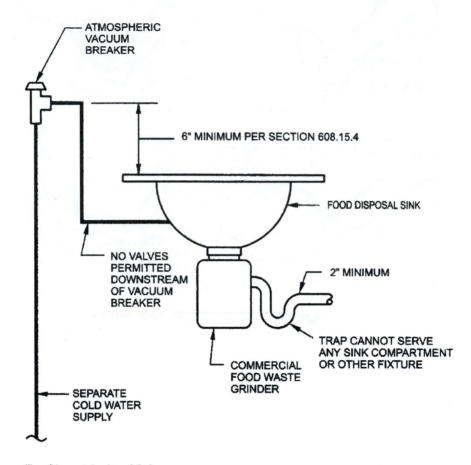

ATMOSPHERIC
VACUUM
BREAKER

6" MINIMUM PER SECTION 608.15.4

FOOD DISPOSAL SINK

NO VALVES
PERMITTED
DOWNSTREAM
OF VACUUM
BREAKER

2" MINIMUM

TRAP CANNOT SERVE
ANY SINK COMPARTMENT
OR OTHER FIXTURE

COMMERCIAL
FOOD WASTE
GRINDER

SEPARATE
COLD WATER
SUPPLY

For SI: 1 inch = 25.4 mm.

FIGURE 4.3 Commercial food waste grinder installation. *Copyright 2002, International Code Council, Inc., Falls Church, Virginia. Reproduced with permission. All rights reserved.*

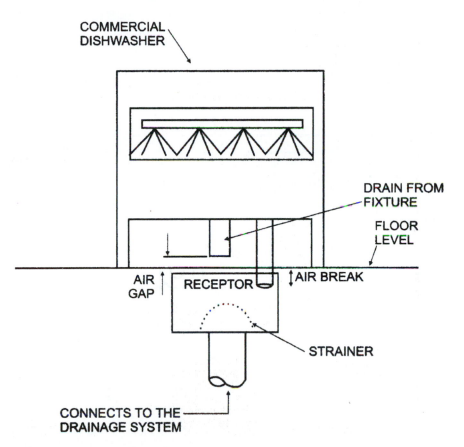

COMMERCIAL
DISHWASHER

DRAIN FROM
FIXTURE

FLOOR
LEVEL

AIR
GAP

RECEPTOR

AIR BREAK

STRAINER

CONNECTS TO THE
DRAINAGE SYSTEM

FIGURE 4.4 Commercial dishwasher waste connection. *Copyright 2002, International Code Council, Inc., Falls Church, Virginia. Reproduced with permission. All rights reserved.*

—NOTES—

Chapter 5
WATER HEATERS

Water heaters are often taken for granted. The code requirements for water heaters are not extensive, but they are important. Through the years, I've seen a number of water heaters installed that did not meet code expectations. Some of the violations were not extremely serious. Others were downright dangerous. I assume that most of the water heaters that violated code requirements were replacemens that no one obtained a permit for. This chapter will provide you with sample testing to prepare you for your licensing exam on the topic of water heaters

MULTIPLE-CHOICE EXAM

1.   With a combination potable water heater and space heater, the potable water must be maintained:

    a.   At a warm temperature   b.   At a cool temperature

    c.   Throughout the system   d.   All of the above

2.   With a combination potable water heater and space heater, if the space heater requires water at temperatures in excess of 140 degrees F, the potable water must be maintained:

    a.   At a temperature not to exceed 150 degrees

    b.   At a temperature not to exceed 140 degrees

    c.   At a temperature not to exceed 130 degrees

    d.   At a temperature not to exceed 120 degrees

3.   All tank-type water heaters require a drain that is located:

    a.   At the top of the tank   b.   At the bottom of the tank

    c.   On the side of the tank   d.   Any of the above

4.   Water heaters must be provided with access for:

    a.   Observation   b.   Maintenance

    c.   Servicing and replacement   d.   All of the above

5.   All water heaters must be:

    a.   Blue and white in color   b.   Certified

    c.   Third-party certified   d.   Installed by a master plumber

6.   Water temperature from a tankless water heater intended for domestic use must not exceed:

    a.   120 degrees   b.   130 degrees

    c.   140 degrees   d.   150 degrees

7. Water heaters installed for domestic use must have the working pressure clearly and indelibly stamped in the metal or:

    a. Marked on a pipe tag

    b. Marked on a plate welded to the tank or otherwise permanently attached

    c. Marked on the top of the tank

    d. Any of the above

8. Pressure markings on water heaters must be:

    a. In an accessible position outside the tank

    b. Accessible for inspection

    c. Accessible for reinspection

    d. All of the above

9. All hot-water supply systems must be equipped with:

    a. Automatic temperature controls

    b. Manual temperature controls

    c. Systematic temperature controls

    d. None of the above

10. Water heaters with an ignition source that are installed in garages shall be elevated such that the source of the ignition is not less than _____ inches above the garage floor:

    a. 12   b. 14

    c. 18   d. 24

11. Water heaters using solid, liquid, or gas fuel must not be installed in a room containing air-handling machinery when such room is used as a:

    a. Plenum   b. Garage

    c. Either of the above   d. Neither of the above

12. A gas-fired water heater that is not a direct-vent model may not be installed in a:

    a. Sleeping room   b. Bathroom

    c. Both A and B   d. Neither A or B

13. A gas-fired water heater that is not a direct-vent model may not be installed in a closet that is accessed through a:

    a. Sleeping room   b. Bathroom

    c. Both A and B   d. Neither A or B

14. Water heaters installed in attics require access. The minimum opening and unobstructed passageway must be large enough to allow the removal of the water heater, and in no case may the height of the access be less than:

a. 24 inches
b. 30 inches
c. 36 inches
d. 60 inches

15. Water heaters installed in attics require access. The minimum opening and unobstructed passageway must be large enough to allow the removal of the water heater, and in no case may the width of the access be less than:

a. 18 inches
b. 22 inches
c. 24 inches
d. 30 inches

16. Water heaters installed in attics require access. The minimum opening and unobstructed passageway must be large enough to allow the removal of the water heater, and in no case may the length of the access be more than:

a. 5 feet
b. 10 feet
c. 20 feet
d. 30 feet

17. Water heaters installed in attics require access. The minimum opening and unobstructed passageway must be large enough to allow the removal of the water heater. The passageway must have a continuous solid flooring that is not less than _____ inches wide:

a. 24
b. 30
c. 36
d. 42

18. Water heaters installed in attics require access. A level service space with minimum dimensions of _____ shall be present at the front or service side of the water heater:

a. 24 inches by 24 inches
b. 30 inches by 30 inches
c. 36 inches by 36 inches
d. 42 inches by 42 inches

19. Water heaters installed in attics require access. The clear access opening dimensions shall be a minimum of _____ where such dimensions are large enough to allow removal of the water heater:

a. 20 inches by 20 inches
b. 20 inches by 30 inches
c. 20 inches by 36 inches
d. 20 inches by 42 inches

20. Valves installed on the cold-water inlet of a water heater must be installed with access on the same _____ as the water heater:

a. Side
b. Floor level
c. Regulations
d. None of the above

21. Valves installed on the cold-water inlet of a water heater must be:

    a.   Stop and waste valves

    b.   Installed near the water heater

    c.   Serving only the water heater

    d.   Both B and C

22. Bottom-fed water heaters must have a:

    a.   Drain pan   b.   Lighted access area

    c.   Vacuum relief valve installed   d.   All of the above

23. The outlet of a relief valve must not be:

    a.   Open

    b.   Connected directly to the drainage system

    c.   Terminated above floor level

    d.   All of the above

24. When water heaters are placed in pans, the depth of the pan must not be less than _____ deep:

    a.   1.5 inches   b.   2 inches

    c.   3 inches   d.   4 inches

! Codealert

The discharge piping serving a pressure relief valve, temperature relief valve or combination must:

1. Not be directly connected to the drainage system.
2. Discharge through an air gap located in the same room as the water heater.
3. Not be smaller than the diameter of the outlet of the valve served and must discharge full size to the air gap.
4. Serve a single relief device and must not connect to piping serving any other relief device or equipment.
5. Discharge to the floor, to an indirect waste receptor or to the outdoors. Where discharging to the outdoors in areas subject to freezing, discharge piping shall be first piped to an indirect waste receptor through an air gap located in a conditioned area.
6. Discharge in a way that does not cause personal or structural injury.
7. Discharge to an end point that can be observed by the building tenants.
8. Not be trapped.
9. Be installed so as to flow by gravity.
10. Not end more than 6 inches above the floor or waste receptor.
11. Not have a threaded connection at the end of the piping.
12. Not have valves or tee fittings.
13. Be constructed of materials listed in section 605.4 or materials tested, rated and approved for use in accordance with ASME A112.4.1.

25.   Water-heater pans must be drained by an indirect waste pipe that has a minimum diameter of:

    a.   1/2 inch   b.   3/4 inch

    c.   1 inch   d.   1.5 inches

26.   When used for domestic purposes, the water temperature from a tankless water heater may not exceed _____.

    a.   120 degrees   b.   100 degrees

    c.   140 degrees   d.   110 degrees

27.   All water heaters need to be _____ certified.

    a.   third-party   b.   accurately

    c.   fourth-party   d.   energy sufficient

28.   When earthquake loads are applicable, water heaters must be installed in accordance with:

    a.   Local building codes   b.   Homeowners Insurance Policy

    c.   Local town office   d.   International Building Code

29.   When planning an exit route for an attic water heater, the minimum height must be:

    a.   20 feet   b.   30 inches

    c.   22 inches   d.   24 feet

30.   Offsets with angle more than 45 degrees are considered to be _____ offsets.

    a.   vertical   b.   terminal

    c.   horizontal   d.   continuous

31.   Every water heater is to be equipped with a drain valve near the _____ of the water heater.

    a.   outside   b.   top

    c.   inside   d.   bottom

32.   Code requires both hot water heaters and hot water storage tanks to be _____.

    a.   accessible   b.   kept covered

    c.   enclosed   d.   installed

33.   Horizontal vent connectors must not be greater than _____ of the vertical height.

    a.   the length   b.   60 degrees

    c.   75 percent   d.   .25 inch

34. Which of the following might a venting system consist of?

   a.   Chimneys   b.   Steel piping

   c.   PVC piping   d.   None of the above

35. Vents must terminate through which of these?

   a.   Walls   b.   Floors

   c.   Chimney   d.   Roofs

This concludes the multiple-choice questions for water heaters. Take the following true-false test and then check your answers.

TRUE-FALSE EXAM

1.   Water heaters must be protected to prevent siphoning of their contents.

   True   False

2.   When water heaters are installed in locations where leakage of the tank or connections will cause damage, the water heater must be installed in a pan.

   True   False

3.   All water heaters must be provided with an approved, self-closing (levered) pressure-relief valve and a temperature-relief valve or a combination thereof.

   True   False

4.   The drain from a water-heater pan may be reduced in its diameter once it is 6 inches beyond the edge of the pan.

   True   False

5.   The discharge pipe from a relief valve must be trapped.

   True   False

6.   Temperature-relief valves must be installed so that they are actuated by the water in the top 12 inches of the tank.

   True   False

7.   Temperature-relief valves must have a temperature setting of not more than 210 degrees F.

   True   False

8.   Pressure-relief valves must have a pressure-rating setting that does not exceed the manufacturer's rated working pressure or 150 PSI, whichever is less.

     True   False

9.   Direct-vent, gas-fired water heaters must not be installed in sleeping room.

     True   False

10. All water heaters must be certified by a third party.

     True   False

11. Water heaters are sometimes used as part of a space heating system.

     True   False

12. Water heaters do not need to be accessible for replacement.

     True   False

13. It doesn't make a difference where the water heater is located.

     True   False

14. The minimum width for the exit route of a water heater is 22 inches.

     True   False

15. Relief valves are optional.

     True   False

16. Energy cutoff valves are required on all water heaters that are automatically controlled.

     True   False

17. The discharge tube from a relief valve must be connected to a drainage system.

     True   False

18. The venting of water heaters that require venting is regulated by electrical provider.

     True   False

19. Safety pans must have a minimum depth of 1.5 inches.

     True   False

20. Draft hoods for water heaters must be located in the same room or space as the combustion air opening for the water heater.

     True   False

21. Every water heater is required to bear a label of an approved agency.

   True   False

22. Potability of water must be maintained at all times.

   True   False

23. A continuous solid floor is required in the exit area, and the flooring must be at least 32 inches wide.

   True   False

24. Unfired hot water storage tanks must be insulated so that heat loss is limited to the maximum of 15 BTUs.

   True   False

25. The term BTU stands for British Thermostat Users.

   True   False

MULTIPLE-CHOICE ANSWERS

1. c	8. d	15. b	22. c	29. b
2. b	9. a	16. c	23. b	30. c
3. b	10. c	17. a	24. a	31. d
4. d	11. a	18. b	25. b	32. a
5. c	12. c	19. b	26. c	33. c
6. c	13. c	20. b	27. a	34. c
7. b	14. b	21. d	28. d	35. d

TRUE-FALSE ANSWERS

1. True	8. True	15. False	21. True
2. True	9. False	16. True	22. True
3. True	10. True	17. False	23. False
4. False	11. False	18. False	24. False
5. False	12. False	19. True	25. False
6. False	13. False	20. True	
7. True	14. True		

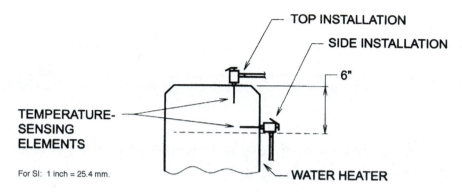

FIGURE 5.1 Temperature and pressure relief valve installation. *Copyright 2002, International Code Council, Inc., Falls Church, Virginia. Reproduced with permission. All rights reserved.*

Chapter 6
WATER SUPPLY AND DISTRIBUTION

With the subject of approved potable-water systems, there is a great deal to learn. Unlike code requirements for healthcare facilities and institutional plumbing, which not all plumbers get involved in, the rules and regulations pertaining to potable water supplies affect all plumbers, even service and repair plumbers. As a plumber, there is no good way to avoid working with potable water supplies.

Potable water is, of course, another name for drinking water. Good, clean water is essential in life, and it is the job of plumbers to install systems that will supply potable water. If plumbers make mistakes in their installation methods or material choices, serious health problems can arise. If you make a mistake that creates a health problem for the person you installed the system for, you could be sued. Your employer could be sued. What's worse, someone could suffer from illness, serious burns, or even death from your mistake. Potable water is not a subject to be taken lightly.

You can rest assured that there will be numerous questions on your licensing exam that pertain to potable water. There will be more than enough questions of this type to cause you great difficulty in passing the exam if you are not well prepared with a depth of knowledge on water distribution.

MULTIPLE-CHOICE EXAM

1. Water that is not safe for drinking is called:

    a. Potable water   b. Nonpotable water

    c. Either A or B   d. Neither A or B

2. Potable water may be used for which of the following purposes:

    a. Personal use   b. Culinary use

    c. Both A and B   d. Neither A or B

3. A valve that is pressure-actuated , held closed by a spring or other means, and designed to relieve pressure automatically at a preset point is called:

    a. A relief valve   b. A pressure-reducing valve

    c. A vacuum breaker   d. None of the above

4. A device used to absorb pressure surges that occur when a flow of water is stopped rapidly is known as:

    a. A pressure-reducing valve   b. A vacuum breaker

    c. A water-hammer arrester   d. None of the above

5. When a valve can be opened to its full-open position and provide a passageway with a diameter equal to that of the pipe in which the valve is installed, the valve is called:

    a. A stop valve   b. A tempering valve

    c. A full-open valve   d. None of the above

6.   What type of valve combines to protect against excessive temperature and pressure build-ups:

   a.   A temperature-and-pressure relief valve

   b.   A temperature-relief valve

   c.   A tempering valve

   d.   None of the above

7.   A valve or faucet that closes automatically when released manually is called a:

   a.   Timed shut-off   b.   A stop valve

   c.   A quick-closing valve   d.   None of the above

8.   A valve or faucet that is controlled by a mechanical means for fast-action closing is called a:

   a.   Supply valve   b.   A stop valve

   c.   A quick-closing valve   d.   None of the above

9.   Potable water must be supplied for which of the following uses:

   a.   Drinking   b.   Bathing

   c.   Culinary uses   d.   All of the above

10.   A pipe that supplies water to branches and extends at least one full story is called:

   a.   A fixture supply   b.   An artery

   c.   A riser   d.   A conductor

11.   The pressure of the water supply near a faucet while the faucet is wide open and flowing at full force is known as:

   a.   Street pressure   b.   Flow pressure

   c.   Static pressure   d.   Controlled pressure

12.   A pipe that runs from a source of potable water to the water-distribution system of a building is known as:

   a.   A water main   b.   A well pipe

   c.   A main   d.   A water service

13.   Which of the following are required to make up a water-supply system:

   a.   Water distribution pipes   b.   A water service pipe

   c.   Both A and B   d.   A check valve

14.   Hot water is required in buildings where:

    a.   People have permanent residences

    b.   Livestock is maintained

    c.   Both A and B

    d.   Either A or B

15.   Hot water is required in buildings where:

    a.   People have permanent residences

    b.   Livestock is maintained

    c.   People are employed and plumbing facilities are available

    d.   Both A and C

16.   Cold water is required in buildings where:

    a.   Plumbing is installed

    b.   People are employed and plumbing facilities are available

    c.   People maintain permanent residences

    d.   All of the above

17.   Potable water is required for which of the following:

    a.   Food processing

    b.   Medical and pharmaceutical processing

    c.   Culinary purposes

    d.   All of the above

18.   The minimum diameter of a water-service pipe is:

    a.   1/2 inch   b.   3/4 inch

    c.   1 inch   d.   1 1/4 inch

19.   The minimum size of a fixture water-supply pipe for a bidet is:

    a.   3/8 inch   b.   1/2 inch

    c.   3/4 inch   d.   1 inch

> **! Code**alert
>
> Where water pressure exceeds 160 psi, piping material must have a minimum rated working pressure equal to the highest available pressure. Water service piping materials not third-party certified for water distribution must terminate at or before the full open valve located at the entrance to the structure. All ductile iron water service piping must be cement mortar lined in accordance with AWWA C104.

20. The minimum size of a fixture water-supply pipe for a bathtub is:

    a. 3/8 inch   b. 1/2 inch

    c. 3/4 inch   d. 1 inch

21. The minimum size of a fixture water-supply pipe for a combination sink and tray is:

    a. 3/8 inch   b. 1/2 inch

    c. 3/4 inch   d. 1 inch

22. The minimum size of a fixture water-supply pipe for a domestic dishwasher is:

    a. 3/8 inch   b. 1/2 inch

    c. 3/4 inch   d. 1 inch

23. The minimum size of a fixture water-supply pipe for a drinking fountain is:

    a. 3/8 inch   b. 1/2 inch

    c. 3/4 inch   d. 1 inch

24. The minimum size of a fixture water-supply pipe for a hose bibb is:

    a. 3/8 inch   b. 1/2 inch

    c. 3/4 inch   d. 1 inch

25. The minimum size of a fixture water-supply pipe for a kitchen sink is:

    a. 3/8 inch   b. 1/2 inch

    c. 3/4 inch   d. 1 inch

26. The minimum size of a fixture water-supply pipe for a laundry tray with up to three compartments is:

    a. 3/8 inch   b. 1/2 inch

    c. 3/4 inch   d. 1 inch

27. The minimum size of a fixture water-supply pipe for a lavatory is:

    a. 3/8 inch   b. 1/2 inch

    c. 3/4 inch   d. 1 inch

28. The minimum size of a fixture water-supply pipe for a single shower head is:

    a. 3/8 inch   b. 1/2 inch

    c. 3/4 inch   d. 1 inch

29.   The minimum size of a fixture water-supply pipe for a service sink is:

a.   3/8 inch   b.   1/2 inch

c.   3/4 inch   d.   1 inch

30.   The minimum size of a fixture water-supply pipe for a flush-tank urinal is:

a.   3/8 inch   b.   1/2 inch

c.   3/4 inch   d.   1 inch

31.   The minimum size of a fixture water-supply pipe for a flush-valve urinal is:

a.   3/8 inch   b.   1/2 inch

c.   3/4 inch   d.   1 inch

32.   The minimum size of a fixture water-supply pipe for a wall hydrant is:

a.   3/8 inch   b.   1/2 inch

c.   3/4 inch   d.   1 inch

33.   The minimum size of a fixture water-supply pipe for a flush-tank toilet is:

a.   3/8 inch   b.   1/2 inch

c.   3/4 inch   d.   1 inch

34.   The minimum size of a fixture water-supply pipe for a flush-valve toilet is:

a.   3/8 inch   b.   1/2 inch

c.   3/4 inch   d.   1 inch

35.   A water-service pipe and a building sewer must be separated by:

a.   A masonry barrier   b.   Undisturbed earth

c.   Compacted earth   d.   Either B or C

36.   The bottom of a water-service pipe should never be closer than _____ inches from the top of a sewer:

a.   6   b.   12

c.   18   d.   24

37.   If a water-service pipe is installed in a common trench with a sewer, the water-service pipe must be placed to one side of the trench and installed on:

a.   An even grade

b.   A solid shelf of earth

c.   A pitch of 1/4 inch per foot

d.   None of the above

38.   Potable-water-service pipes may not be installed in, under, or above:

a.   Cesspools

b.   Septic tanks

c.   Septic discharge fields

d.   All of the above

39.   Caution must be exercised and devices may be required to prevent the contents of hot- and cold-water pipes from:

a.   Flowing between the two piping systems

b.   Losing their ambient temperature

c.   Both A and B

d.   Neither A or B

40.   Fixture supply pipes must not terminate at distances greater than _____ inches from the point of connection with the fixture. Fill in the blank:

a.   12

b.   16

c.   30

d.   32

41.   Fixture supply pipes must extend to the floor or _____ adjacent to the fixture:

a.   The next fixture

b.   Wall

c.   Ceiling

d.   Any of the above

42.   An approved pressure-reducing valve must be installed in a water-service pipe of a building near where the pipe enters a building when the street pressure of the water main providing potable water exceeds _____ pounds per square inch. Fill in the blank:

a.   60

b.   75

c.   80

d.   100

43.   Pressure-reducing valves, as described in question 42, must reduce the water pressure entering the water-distribution system to a point not to exceed _____ pounds per square inch. Fill in the blank:

a.   60

b.   75

c.   80

d.   100

! Codealert

The maximum water consumption flow rates and quantities for all plumbing fixtures and fixture filling must be in accordance with Table 6.4. There are two exceptions:
1. Blowout design water closets having a maximum water consumption of 3.5 gallons per flushing cycle.
2. Clinical sinks having a maximum water consumption of 4.5 gallons.

44. Under normal conditions, a water hammer arrester is required wherever certain valves are used. Which of the following valves are of a type to require such a device:

a. Gate valves   b. Stop-and-waste valves

c. Quick-closing valves   d. All of the above

45. Water-hammer arresters are required to be installed so that they are:

a. In compliance with ASSE 1010

b. Readily accessible

c. Are not concealed in any way

d. Easy to clean

46. When the pressure of a water main that supplies potable water to a building fluctuates, the water-distribution system in the building must be designed for the _____ pressure available. Fill in the blank:

a. Minimum   b. Maximum

c. Average   d. Consolidated

47. When manifolds are used for the distribution of potable water, the cutoff valves located at the manifold must be:

a. Full-open valves

b. Full-flow valves

c. Gate valves

d. Labeled to identify the fixture being served by the valve

48. Where manifolds are installed, they must be:

a. Accessible

b. Readily accessible

c. 42 inches above the finished floor

d. 60 inches above the finished floor

49. The maximum amount of water consumption for toilets in a majority of locations shall be no more than _____ gallons per flushing cycle:

a. 1.5   b. 1.6

c. 3.5   d. 4

50. The maximum amount of water consumption for urinals in a majority of locations shall be no more than _____ gallons per flushing cycle:

a. 1   b. 1.5

c. 1.6   d. 2.5

51. When the water pressure is 60 pounds per square inch, the maximum flow rate for a single shower head shall not exceed _____ gallons per minute. Fill in the blank:

    a. 1.5

    b. 1.75

    c. 2

    d. 2.5

52. When the water pressure is 60 pounds per square inch, the maximum flow rate for a public lavatory shall not exceed _____ gallons per minute. Fill in the blank:

    a. 0.5

    b. 1

    c. 1.25

    d. 1.5

53. When the water pressure is 60 pounds per square inch, the maximum flow rate for a private lavatory shall not exceed _____ gallons per minute. Fill in the blank:

    a. 1.5

    b. 1.75

    c. 2

    d. 2.2

54. When the water pressure is 60 pounds per square inch, the maximum flow rate for a sink faucet shall not exceed _____ gallons per minute. Fill in the blank:

    a. 1.5

    b. 1.75

    c. 2

    d. 2.2

55. The maximum flow rate for a public lavatory with self-closing valves shall not exceed _____ gallons per minute. Fill in the blank:

    a. 0.25

    b. 0.5

    c. 0.75

    d. 1

56. What type of valve is required near the curb when a building water-service pipe connects to a public water supply:

    a. Stop-and-waste valve

    b. Stop valve

    c. Full-open valve

    d. Compression valve

57. When a water service becomes a water-distribution main, which of the following types of valves is required:

    a. Stop-and-waste valve

    b. Stop valve

    c. Full-open valve

    d. Compression valve

58. A full-open valve is required on the discharge side of every _____. Fill in the blank:

    a. Fixture

    b. Sink

    c. Water meter

    d. Water heater

> **! Code**alert
>
> All valves must be of an approved type and compatible with the type of piping installed in the system. Ball valves, gate vales, globe valves, and plug valves used to supply drinking water must meet the requirements of NSF 61.

59. A pipe that provides a water supply to a gravity or pressurized water tank is required to be fitted with _____. Fill in the blank:

    a. A stop-and-waste valve   b. A stop valve

    c. A full-open valve   d. A compression valve

60. A full-open valve is required on the supply side of every _____. Fill in the blank:

    a. Fixture   b. Sink

    c. Dishwasher   d. Water heater

61. Which of the following cutoff valves may be used with a sillcock:

    a. Stop-and-waste valve   b. Full-open valve

    c. Gate valve   d. Any of the above

62. A cutoff valve is required on a pipe that supplies water to which of the following:

    a. Appliances   b. Mechanical equipment

    c. Sillcocks   d. All of the above

63. When valves are installed in locations that are not adjacent to the fixture or appliance being served, the valves must be _____. Fill in the blank:

    a. Full-open valves

    b. Full-flow valves

    c. Identified and labeled to show the fixture being served

    d. Both A and C

64. Service and hose-bibb valves must be _____. Fill in the blank:

    a. Full-open valves

    b. Full-flow valves

    c. Identified and labeled to show the fixture being served

    d. Both A and C

65. Combination stop-and-waste valves are not allowed to be installed _____. Fill in the blank:
    a. In finished walls
    b. In ceilings
    c. Underground
    d. All of the above

66. The main water pipe delivering potable water to a building is called:
    a. A water service pipe
    b. A drainage pipe
    c. An ABS pipe
    d. Ductile iron water pipe

67. Fixture supplies are the tubes or pipes that rise from the _____ branch
    a. pipe
    b. fixture
    c. pressure
    d. water

68. A fixture supply may not have a length of more than _____ inches.
    a. 12
    b. 15
    c. 25
    d. 30

69. Banging pipes are normally the results of:
    a. Unruly kids
    b. Bent or crooked pipes
    c. Water hammer
    d. Being installed wrong

70. You may not use galvanized straphanger to support _____ pipe.
    a. metal
    b. copper
    c. iron
    d. plastic

71. Public lavatories equipped with a self-closing faucet may produce up to _____ per use.
    a. .25
    b. 30
    c. .050
    d. .50

72. Which of the following require an air gap as protection from backflow?
    a. Boilers
    b. Sinks
    c. Bathtubs
    d. Both b and c

73. The standard working pressure for a water heater is:
    a. 112 psi.
    b. 120 psi.
    c. 140 psi.
    d. 125 psi.

> ! **Code**alert
>
> A dual check-valve backflow preventor installed on the water supply system, must comply with ASSE 1024 or CSA B64.6. Plumbing fixture fittings must provide backflow protection in accordance with ASME A112.18.1.

74. _____ is required around the exterior casing of drilled and driven wells.

a. Cement

b. Grouting

c. Dirt

d. Sand

75. Maintaining energy efficiency must conform to the:

a. International Emergency Construction Code

b. International Conservation Code

c International Energy Conservation Code

d. The local power service provider

This concludes the multiple-choice testing on potable water systems. We are about to begin the true-false testing on this subject. When you are ready, proceed to the true-false exam.

TRUE-FALSE EXAM

1. It is a violation of the plumbing code to make a connection between a potable water- supply pipe and any other pipe that is not used solely for the purpose of conveying potable water unless suitable backflow prevention is provided.

True   False

2. It is a violation of the plumbing code to connect a private water supply to a public water supply without prior approval from code officials.

True   False

3. An atmospheric vacuum breaker consists of a body, a checking member, and an atmospheric opening.

True   False

4. A double check-valve backflow-prevention assembly is made up of two independently acting internally loaded check valves, two properly located test cocks, and two isolation valves.

True   False

5.   Soldered joints associated with potable water pipe and fittings must contain no more than 8 percent lead.

True   False

6.   All required shutoff valves must be readily accessible.

True   False

7.   When the water pressure of a potable water system is inadequate, a tank and pump must be used to boost the pressure.

True   False

8.   Relief valves located within a building must be trapped.

True   False

9.   The terminal end of a drain from a relief valve must be provided with threads in the event that a hose connection is needed.

True   False

10.   The minimum amount of ground cover allowed over a water-service yard piping system is 12 inches.

True   False

11.   A water service pipe that shares a trench with a sewer must be separated from the drainage pipe.

True   False

12.   The minimum size of water supply pipes for a kitchen sink is 1-inch.

True   False

13.   All flexible water connectors must be accessible.

True   False

14.   Booster pumps are required to be equipped with low-water cutoffs.

True   False

15.   A piece of wire is acceptable to use as a hanger.

True   False

16.   Every sill cock must be equipped with an individual relief vent.

True   False

17.   It is not acceptable to install stop-and-waste valves above ground.

True   False

18. Private water supplies and pump suction lines must have a distance of 100 feet from a pasture.
True   False

19. Cross-connections are prohibited, except where approved protective devices are installed.
True   False

20. Traps need to be installed on the discharge piping from safety pans.
True   False

21. Part of a plumber's responsibility is to provide safe drinking water.
True   False

22. The only fixture that must have potable water available is a toilet.
True   False

23. A water service pipe must have a diameter of at least .75 inch.
True   False

24. It is permissible to use potable water for flushing toilets and urinals.
True   False

25. There is not much of a difference between non-potable and potable water.
True   False

MULTIPLE-CHOICE ANSWERS

1. b	16. d	31. c	46. b	61. d
2. c	17. d	32. b	47. d	62. d
3. a	18. b	33. a	48. a	63. c
4. c	19. a	34. d	49. b	64. c
5. c	20. b	35. d	50. a	65. c
6. a	21. b	36. b	51. d	66. a
7. c	22. b	37. b	52. a	67. b
8. c	23. a	38. d	53. d	68. d
9. d	24. b	39. a	54. d	69. c
10. c	25. b	40. c	55. a	70. b
11. b	26. b	41. b	56. c	71. a
12. d	27. a	42. c	57. c	72. d
13. c	28. b	43. c	58. c	73. d
14. a	29. b	44. c	59. c	74. b
15. d	30. b	45. a	60. d	75. a

TRUE-FALSE ANSWERS

1. True	8. False	14. True	20. False
2. True	9. False	15. False	21. True
3. True	10. True	16. False	22. False
4. True	11. True	17. False	23. True
5. True	12. False	18. True	24. True
6. False	13. True	19. True	25. False
7. True			

TABLE 6.1 Water service pipe.

MATERIAL	STANDARD
Acrylonitrile butadiene styrene (ABS) plastic pipe	ASTM D 1527; ASTM D 2282
Asbestos-cement pipe	ASTM C 296
Brass pipe	ASTM B 43
Chlorinated polyvinyl chloride (CPVC) plastic pipe	ASTM D 2846; ASTM F 441; ASTM F 442; CSA B137.6
Copper or copper-alloy pipe	ASTM B 42; ASTM B 302
Copper or copper-alloy tubing (Type K, WK, L, WL, M or WM)	ASTM B 75; ASTM B 88; ASTM B 251; ASTM B 447
Cross-linked polyethylene (PEX) plastic tubing	ASTM F 876; ASTM F 877; CSA B137.5
Cross-linked polyethylene/aluminum/cross-linked polyethylene (PEX-AL-PEX) pipe	ASTM F 1281; CSA B137.10M
Cross-linked polyethylene/aluminum/high-density polyethylene (PEX-AL-HDPE)	ASTM F 1986
Ductile iron water pipe	AWWA C151; AWWA C115
Galvanized steel pipe	ASTM A 53
Polybutylene (PB) plastic pipe and tubing	ASTM D 2662; ASTM D 2666; ASTM D 3309; CSA B137.8M
Polyethylene (PE) plastic pipe	ASTM D 2239; CSA B137.1
Polyethylene (PE) plastic tubing	ASTM D 2737; CSA B137.1
Polyethylene/aluminum/polethylene (PE-AL-PE) pipe	ASTM F 1282; CSA B137.9
Polypropylene (PP) plastic pipe or tubing	ASTM F 2389; CSA B137.11
Polyvinyl chloride (PVC) plastic pipe	ASTM D 1785; ASTM D 2241; ASTM D 2672; CSA B137.3
Stainless steel pipe (Type 304/304L)	ASTM A 312; ASTM A 778
Stainless steel pipe (Type 316/316L)	ASTM A 312; ASTM A 778

TABLE 6.2   Minimum sizes of fixture water supply pipes.
Copyright 2006, International Code Council, Inc., Falls Church, Virginia.
Reproduced with permission. All rights reserved.

FIXTURE	MINIMUM PIPE SIZE (Inch)
Bathtubs[a] (60″ × 32″ and smaller)	1/2
Bathtubs[a] (larger than 60″ × 32″)	1/2
Bidet	3/8
Combination sink and tray	1/2
Dishwasher, domestic[a]	1/2
Drinking fountain	3/8
Hose bibbs	1/2
Kitchen sink[a]	1/2
Laundry, 1, 2 or 3 compartments[a]	1/2
Lavatory	3/8
Shower, single head[a]	1/2
Sinks, flushing rim	3/4
Sinks, service	1/2
Urinal, flush tank	1/2
Urinal, flush valve	3/4
Wall hydrant	1/2
Water closet, flush tank	3/8
Water closet, flush valve	1
Water closet, flushometer tank	3/8
Water closet, one piece[a]	1/2

For SI: 1 inch = 25.4 mm, 1 foot = 304.8 mm,
          1 pound per square inch = 6.895 kPa.

a. Where the developed length of the distribution line is 60 feet or less, and the available pressure at the meter is a minimum of 35 psi, the minimum size of an individual distribution line supplied from a manifold and installed as part of a parallel water distribution system shall be one nominal tube size smaller than the sizes indicated.

TABLE 6.3   Water distribution system design criteria required capacity at fixture supply pipe outlets. *Copyright 2006, International Code Council, Inc., Falls Church, Virginia. Reproduced with permission. All rights reserved.*

FIXTURE SUPPLY OUTLET SERVING	FLOW RATE (gpm)	FLOW PRESSURE (psi)
Bathtub	4	8
Bidet	2	4
Combination fixture	4	8
Dishwasher, residential	2.75	8
Drinking fountain	0.75	8
Laundry tray	4	8
Lavatory	2	8
Shower	3	8
Shower, temperature controlled	3	20
Sillcock, hose bibb	5	8
Sink, residential	2.5	8
Sink, service	3	8
Urinal, valve	15	15
Water closet, blow out, flushometer valve	35	25
Water closet, flushometer tank	1.6	15
Water closet, siphonic, flushometer valve	25	15
Water closet, tank, close coupled	3	8
Water closet, tank, one piece	6	20

For SI: 1 pound per square inch = 6.895 kPa, 1 gallon per minute = 3.785 L/m.

TABLE 6.4   Maximum flow rates and consumption for plumbing fixtures and fixture fittings. *Copyright 2006, International Code Council, Inc., Falls Church, Virginia. Reproduced with permission. All rights reserved.*

PLUMBING FIXTURE OR FIXTURE FITTING	MAXIMUM FLOW RATE OR QUANTITY[b]
Lavatory, private	2.2 gpm at 60 psi
Lavatory, public, (metering)	0.25 gallon per metering cycle
Lavatory, public (other than metering)	0.5 gpm at 60 psi
Shower head[a]	2.5 gpm at 80 psi
Sink faucet	2.2 gpm at 60 psi
Urinal	1.0 gallon per flushing cycle
Water closet	1.6 gallons per flushing cycle

For SI: 1 gallon = 3.785 L, 1 gallon per minute = 3.785 L/m,
1 pound per square inch = 6.895 kPa.
a. A hand-held shower spray is a shower head.
b. Consumption tolerances shall be determined from referenced standards.

TABLE 6.5 Application of backflow preventers.

DEVICE	DEGREE OF HAZARD	APPLICATION	APPLICABLE STANDARDS
Air gap	High or low hazard	Backsiphonage or backpressure	ASME A112.1.2
Air gap fittings for use with plumbing fixtures, appliances and appurtenances	High or low hazard	Backsiphonage or backpressure	ASME A112.1.3
Antisiphon-type fill valves for gravity water closet flush tanks	High hazard	Backsiphonage only	ASSE 1002, CSA B125
Backflow preventer for carbonated beverage machines	Low hazard	Backpressure or backsiphonage Sizes $1/4" - 3/8"$	ASSE 1022, CSA B64.3.1
Backflow preventer with intermediate atmospheric vents	Low hazard	Backpressure or backsiphonage Sizes $1/4" - 3/4"$	ASSE 1012, CSA B64.3
Barometric loop	High or low hazard	Backsiphonage only	(See Section 608.13.4)
Double check backflow prevention assembly and double check fire protection backflow prevention assembly	Low hazard	Backpressure or backsiphonage Sizes $3/8" - 16"$	ASSE 1015, AWWA C510, CSA B64.5, CSA B64.5.1
Double check detector fire protection backflow prevention assemblies	Low hazard	Backpressure or backsiphonage (Fire sprinkler systems) Sizes 2" - 16"	ASSE 1048
Dual-check-valve-type backflow preventer	Low hazard	Backpressure or backsiphonage Sizes $1/4" - 1"$	ASSE 1024, CSA B64.6
Hose connection backflow preventer	High or low hazard	Low head backpressure, rated working pressure, backpressure or backsiphonage Sizes $1/2"-1"$	ASSE 1052, CSA B64.2.1.1

Device	Degree of hazard	Applicable pressure and sizes	Standard reference
Hose connection vacuum breaker	High or low hazard	Low head backpressure or backsiphonage Sizes $1/2''$, $3/4''$, $1''$	ASSE 1011, CSA B64.2, CSA B64.2.1
Laboratory faucet backflow preventer	High or low hazard	Low head backpressure and backsiphonage	ASSE 1035, CSA B64.7
Pipe-applied atmospheric-type vacuum breaker	High or low hazard	Backsiphonage only Sizes $1/4'' - 4''$	ASSE 1001, CSA B64.1.1
Pressure vacuum breaker assembly	High or low hazard	Backsiphonage only Sizes $1/2'' - 2''$	ASSE 1020, CSA B64.1.2
Reduced pressure principle backflow preventer and reduced pressure principle fire protection backflow preventer	High or low hazard	Backpressure or backsiphonage Sizes $3/8'' - 16''$	ASSE 1013, AWWA C511, CSA B64.4, CSA B64.4.1
Reduced pressure detector fire protection backflow prevention assemblies	High or low hazard	Backsiphonage or backpressure (Fire sprinkler systems)	ASSE 1047
Spillproof vacuum breaker	High or low hazard	Backsiphonage only Sizes $1/4'' -2''$	ASSE 1056
Vacuum breaker wall hydrants, frost-resistant, automatic draining type	High or low hazard	Low head backpressure or backsiphonage Sizes $3/4''$, $1''$	ASSE 1019, CSA B64.2.2

For SI: 1 inch = 25.4 mm.

TABLE 6.6 Minimum required air gaps. *Copyright 2006, International Code Council, Inc., Falls Church, Virginia. Reproduced with permission. All rights reserved.*

FIXTURE	MINIMUM AIR GAP	
	Away from a wall[a] (inches)	Close to a wall (inches)
Lavatories and other fixtures with effective opening not greater than $^1/_2$ inch in diameter	1	$1^1/_2$
Sink, laundry trays, gooseneck back faucets and other fixtures with effective openings not greater than $^3/_4$ inch in diameter	1.5	2.5
Over-rim bath fillers and other fixtures with effective openings not greater than 1 inch in diameter	2	3
Drinking water fountains, single orifice not greater than $^7/_{16}$ inch in diameter or multiple orifices with a total area of 0.150 square inch (area of circle $^7/_{16}$ inch in diameter)	1	$1^1/_2$
Effective openings greater than 1 inch	Two times the diameter of the effective opening	Three times the diameter of the effective opening

For SI: 1 inch = 25.4 mm.

a. Applicable where walls or obstructions are spaced from the nearest inside-edge of the spout opening a distance greater than three times the diameter of the effective opening for a single wall, or a distance greater than four times the diameter of the effective opening for two intersecting walls.

TABLE 6.7 Distance from contamination to private water supplies and pump suction lines. *Copyright 2006, International Code Council, Inc., Falls Church, Virginia. Reproduced with permission. All rights reserved.*

SOURCE OF CONTAMINATION	DISTANCE (feet)
Barnyard	100
Farm silo	25
Pasture	100
Pumphouse floor drain of cast iron draining to ground surface	2
Seepage pits	50
Septic tank	25
Sewer	10
Subsurface disposal fields	50
Subsurface pits	50

For SI: 1 foot = 304.8 mm.

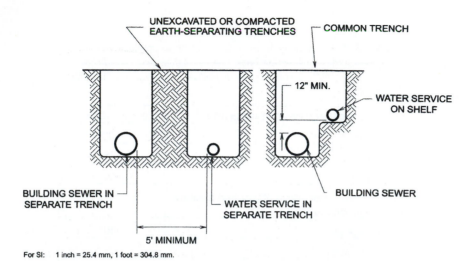

FIGURE 6.1 Separation of water service and building sewer. *Copyright 2002, International Code Council, Inc., Falls Church, Virginia. Reproduced with permission. All rights reserved.*

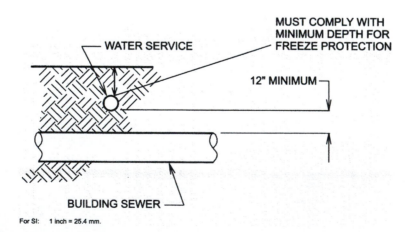

FIGURE 6.2 Water service crossing over building sewer. *Copyright 2002, International Code Council, Inc., Falls Church, Virginia. Reproduced with permission. All rights reserved.*

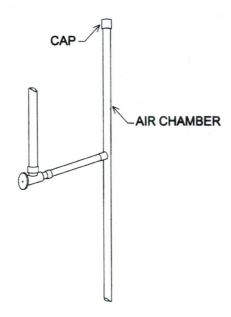

FIGURE 6.3 Example of the piping used to make an air chamber. *Copyright 2002, International Code Council, Inc., Falls Church, Virginia. Reproduced with permission. All rights reserved.*

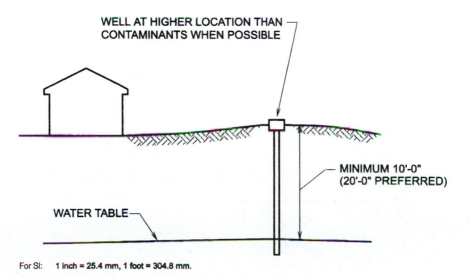

For SI: 1 inch = 25.4 mm, 1 foot = 304.8 mm.

FIGURE 6.4 Depth of well. *Copyright 2002, International Code Council, Inc., Falls Church, Virginia. Reproduced with permission. All rights reserved.*

Chapter 7
SANITARY DRAINAGE SYSTEMS

When installing sanitary drainage systems, there is a lot to learn. For example, some types of fittings can be used to turn a pipe from a horizontal run to a vertical run, but not from a horizontal to a horizontal run. The fact is that the rules and regulations pertaining to sanitary drainage systems are extensive. Before you can consider yourself competent to take full charge of a drainage system, you must know how to size drainage pipes and traps. This work is not difficult once you understand the basic principles, but it can be confusing to someone who is not accustomed to doing it.

From a casual reading of the drainage chapter in your code book, you might feel that the chapter is short, concise, and easy to work with. This can be true if you have the foundation of knowledge needed to use the information provided in the code, but if you are not fluent in fixture units and types of fittings allowed for changes in direction, your opinion may change quickly.

The charts and tables that are provided in the code book to identify fixture-unit ratings don't take up much space. They are just a bunch of numbers filling out a table. But will you know how many fixture units a particular type of drain can carry on the day of your exam? You'd better if you want a high score on your exam.

Don't be fooled by the brevity of the chapter in your code book that deals with drainage systems. While the words may be few, the rules and regulations are many. You will also be wise to learn the chapter thoroughly, since much of your licensing exam is likely to be based on drainage systems.

I've talked earlier about how plumbing exams often use creative manipulation of words to confuse the individuals taking the tests. To simulate what I believe you may encounter on your real exam, I will deploy these same tactics in the mock exams you are about to take. Pay attention, take your time, and see how you do.

MULTIPLE-CHOICE EXAM

1. When a drainage system is installed to convey chemical waste, the system must be:
   a. Installed with Schedule-40 plastic pipe
   b. Tested at a pressure of 20 psi
   c. Completely separated from the sanitary drainage system
   d. All of the above

2. When pipe sizing for a sanitary drainage system is computed, the primary factor that determines the appropriate size is:
   a. The fixture-unit load
   b. The slope of the drain
   c. The type of pipe being used for the drain
   d. All of the above

3.   Which of the following must be added into the fixture-unit load when sizing drainage pipe:

a.   Potential future fixtures

b.   Future fixtures that are on the blueprints

c.   Future fixtures that are roughed in

d.   All of the above

4.   For the purposes of sizing a drainage stack, horizontal offsets must be classified as:

a.   Fixture loads

b.   Piping obstructions

c.   Future fixture loads

d.   Part of the stack fixture load

5.   A plumbing fixture that has a trap with a diameter of 1 1/4 inches or less is rated at how many fixture units:

a.   1   b.   2

c.   3   d.   4

6.   A plumbing fixture that has a trap with a diameter of 1 1/2 inches is rated at how many fixture units:

a.   1   b.   2

c.   3   d.   4

7.   A plumbing fixture that has a trap with a diameter of 2 inches is rated at how many fixture units:

a.   1   b.   2

c.   3   d.   4

8.   A plumbing fixture that has a trap with a diameter of 2 1/2 inches is rated at how many fixture units:

a.   1   b.   2

c.   3   d.   4

9.   A plumbing fixture that has a trap with a diameter of 3 inches is rated at how many fixture units:

a.   4   b.   5

c.   6   d.   7

10.   A plumbing fixture that has a trap with a diameter of 4 inches is rated at how many fixture units:

a.   4   b.   5

c.   6   d.   7

11.   Which of the following numbers indicates the minimum slope, per inch per foot, that is required for a
      pipe with a diameter of 2 1/2 inches or less:

      a.   1/16 inch per foot   b.   1/8 inch per foot

      c.   1/4 inch per foot   d.   1/2 inch per foot

12.   Which of the following numbers indicates the minimum slope, per inch per foot, that is required for a
      pipe with a diameter of between 3 and 6 inches:

      a.   1/16 inch per foot   b.   1/8 inch per foot

      c.   1/4 inch per foot   d.   1/2 inch per foot

13.   Which of the following numbers indicates the minimum slope, per inch per foot, that is required for a
      pipe with a diameter of 8 inches or more:

      a.   1/16 inch per foot   b.   1/8 inch per foot

      c.   1/4 inch per foot   d.   1/2 inch per foot

14.   What is the drainage fixture-unit value of a commercial automatic clothes-washer standpipe:

      a.   1 fixture unit   b.   2 fixture units

      c.   3 fixture units   d.   4 fixture units

15.   What is the drainage fixture-unit value of a domestic automatic clothes-washer standpipe:

      a.   1 fixture unit   b.   2 fixture units

      c.   3 fixture units   d.   4 fixture units

16.   What is the drainage fixture-unit value of a bathroom group with a water closet that discharges at a rate
      of 1.6 gallons of water per flush:

      a.   3 fixture units   b.   4 fixture units

      c.   5 fixture units   d.   6 fixture units

17.   What is the drainage fixture-unit value of a domestic bathtub:

      a.   1 fixture unit   b.   2 fixture units

      c.   3 fixture units   d.   4 fixture units

18.   What is the drainage fixture-unit value of a domestic bidet:

      a.   1 fixture unit   b.   2 fixture units

      c.   3 fixture units   d.   4 fixture units

19.   What is the drainage fixture-unit value of a combination sink and tray:

      a.   1 fixture unit   b.   2 fixture units

      c.   3 fixture units   d.   4 fixture units

20.   What is the drainage fixture-unit value of a dental unit:

    a.   1 fixture unit   b.   2 fixture units

    c.   3 fixture units   d.   4 fixture units

21.   What is the drainage fixture-unit value of a dishwasher:

    a.   1 fixture unit   b.   2 fixture units

    c.   3 fixture units   d.   4 fixture units

22.   What is the drainage fixture-unit value of a drinking fountain:

    a.   1/3 fixture unit   b.   1/2 fixture unit

    c.   1 fixture unit   d.   1 1/2 fixture units

23.   What is the drainage fixture-unit value of a floor drain:

    a.   1 fixture unit   b.   2 fixture units

    c.   3 fixture units   d.   4 fixture units

24.   What is the drainage fixture-unit value of a kitchen sink:

    a.   1 fixture unit   b.   2 fixture units

    c.   3 fixture units   d.   4 fixture units

25.   What is the drainage fixture-unit value of a laundry tray:

    a.   1 fixture unit   b.   2 fixture units

    c.   3 fixture units   d.   4 fixture units

26.   What is the drainage fixture-unit value of a lavatory:

    a.   1 fixture unit   b.   2 fixture units

    c.   3 fixture units   d.   4 fixture units

27.   What is the drainage fixture-unit value of a single showerhead:

    a.   1 fixture unit   b.   2 fixture units

    c.   3 fixture units   d.   4 fixture units

28.   What is the drainage fixture-unit value of a sink:

    a.   1 fixture unit   b.   2 fixture units

    c.   3 fixture units   d.   4 fixture units

29.   What is the drainage fixture-unit value of a urinal:

    a.   1 fixture unit   b.   2 fixture units

    c.   3 fixture units   d.   4 fixture units

30.   What is the drainage fixture-unit value of a private water closet that flushes at a rate of 1.6 gallons per minute:

    a.   1 fixture unit   b.   2 fixture units

    c.   3 fixture units   d.   4 fixture units

31.   What is the drainage fixture-unit value of a public water closet that flushes at a rate of 1.6 gallons per minute:

    a.   3 fixture units   b.   4 fixture units

    c.   5 fixture units   d.   6 fixture units

32.   What is the minimum trap size required for a commercial automatic clothes washer:

    a.   1 1/4 inches   b.   1 1/2 inches

    c.   2 inches   d.   3 inch

33.   What is the minimum trap size required for a domestic automatic clothes washer:

    a.   1 1/4 inches   b.   1 1/2 inches

    c.   2 inches   d.   3 inches

34.   What is the minimum trap size required for a bathtub:

    a.   1 1/4 inches   b.   1 1/2 inches

    c.   2 inches   d.   3 inches

35.   What is the minimum trap size required for a bidet:

    a.   1 1/4 inches   b.   1 1/2 inches

    c.   2 inches   d.   3 inches

36.   What is the minimum trap size required for a combination sink and tray:

    a.   1 1/4 inches   b.   1 1/2 inches

    c.   2 inches   d.   3 inches

37.   What is the minimum trap size required for a dental unit:

    a.   1 1/4 inches   b.   1 1/2 inches

    c.   2 inches   d.   3 inches

38. What is the minimum trap size required for a dishwasher:

    a. 1 1/4 inches   b. 1 1/2 inches

    c. 2 inches   d. 3 inches

39. What is the minimum trap size required for a drinking fountain:

    a. 1 1/4 inches   b. 1 1/2 inches

    c. 2 inches   d. 3 inches

40. What is the minimum trap size required for an emergency floor drain:

    a. 1 1/4 inches   b. 1 1/2 inches

    c. 2 inches   d. 3 inches

41. What is the minimum trap size required for a floor drain:

    a. 1 1/4 inches   b. 1 1/2 inches

    c. 2 inches   d. 3 inches

42. What is the minimum trap size required for a kitchen sink:

    a. 1 1/4 inches   b. 1 1/2 inches

    c. 2 inches   d. 3 inches

43. What is the minimum trap size required for a laundry tray:

    a. 1 1/4 inches   b. 1 1/2 inches

    c. 2 inches   d. 3 inches

44. What is the minimum trap size required for a lavatory:

    a. 1 1/4 inches   b. 1 1/2 inches

    c. 2 inches   d. 3 inches

45. What is the minimum trap size required for a single shower head:

    a. 1 1/4 inches   b. 1 1/2 inches

    c. 2 inches   d. 3 inches

46. What is the minimum trap size required for a sink:

    a. 1 1/4 inches   b. 1 1/2 inches

    c. 2 inches   d. 3 inches

47.   A drainpipe that has a diameter of 1 1/4 inches and a fall of 1/4 inch per foot can carry how many fixture units:

   a.   1   b.   1 1/2
   c.   2   d.   3

48.   A drainpipe that has a diameter of 1 1/4 inches and a fall of 1/2 inch per foot can carry how many fixture units:

   a.   1   b.   1 1/2
   c.   2   d.   3

49.   A drainpipe that has a diameter of 1 1/2 inches and a fall of 1/4 inch per foot can carry how many fixture units:

   a.   1   b.   1 1/2
   c.   2   d.   3

50.   A drainpipe that has a diameter of 1 1/2 inches and a fall of 1/2 inch per foot can carry how many fixture units:

   a.   1   b.   1 1/2
   c.   2   d.   3

51.   A drainpipe that has a diameter of 2 inches and a fall of 1/4 inch per foot can carry how many fixture units:

   a.   10   b.   20
   c.   21   d.   26

52.   A drainpipe that has a diameter of 2 inches and a fall of 1/2 inch per foot can carry how many fixture units:

   a.   10   b.   20
   c.   21   d.   26

53.   A drainpipe that has a diameter of 3 inches and a fall of 1/8 inch per foot can carry how many fixture units:

   a.   26   b.   36
   c.   42   d.   50

54.   A drainpipe that has a diameter of 3 inches and a fall of 1/4 inch per foot can carry how many fixture units:

   a.   26   b.   36
   c.   42   d.   50

55. A drainpipe that has a diameter of 3 inches and a fall of 1/2 inch per foot can carry how many fixture units:

    a. 26
    b. 36
    c. 42
    d. 50

56. A drainpipe that has a diameter of 4 inches and a fall of 1/8 inch per foot can carry how many fixture units:

    a. 160
    b. 180
    c. 216
    d. 250

57. A drainpipe that has a diameter of 4 inches and a fall of 1/4 inch per foot can carry how many fixture units:

    a. 160
    b. 180
    c. 216
    d. 250

58. A drainpipe that has a diameter of 4 inches and a fall of 1/2 inch per foot can carry how many fixture units:

    a. 216
    b. 250
    c. 276
    d. 295

59. Which of the following fittings can be used for a change in direction in drainage piping from horizontal to vertical:

    a. one-sixteenth bend
    b. one-eighth bend
    c. one-sixth bend
    d. All of the above

60. Which of the following fittings can be used for a change in direction in drainage piping from horizontal to vertical:

    a. Quarter bend
    b. Short sweep
    c. Long sweep
    d. All of the above

61. Which of the following fittings can be used for a change in direction in drainage piping from horizontal to vertical:

    A. Quarter bend
    B. Wye
    C. Combination wye and one-eighth bend
    D. All of the above

62.   Which of the following fittings can be used for a change in direction in drainage piping from horizontal to vertical only in certain circumstances:

   A. Quarter bend

   B. Wye

   C. Combination wye and one-eighth bend

   D. Sanitary tee

63.   Which of the following fittings may not be used for a change in direction in drainage piping from a vertical to horizontal direction:

   A. Quarter bend

   B. Wye

   C. Combination wye and one-eighth bend

   D. Sanitary tee

64.   Which of the following fittings may be used for a change in direction in drainage piping from a vertical to horizontal direction:

   a. one-sixteenth bend   b.   Sanitary tee

   c. Side-inlet quarter bend   d.   All of the above

65.   Which of the following fittings may be used for a change in direction in drainage piping from a vertical to horizontal direction:

   a. one-eighth bend   b.   one-sixth bend

   c. Wye   d.   All of the above

66.   Which of the following fittings may be used for a change in direction in drainage piping from a vertical to horizontal direction:

   a. one-eighth bend

   b. Combination wye and one-eighth bend

   c. Wye

   d. All of the above

67.   Which of the following fittings may be used for a change in direction in drainage piping from a vertical to horizontal direction when the pipe is at least 3 inches in diameter:

   a. Short sweep

   b. Combination wye and one-eighth bend

   c. Wye

   d. Quarter bend

68. All drainage pipe installed for future fixtures must terminate with a:

    a. Plug
                                b. Cap

    c. Either A or B
                              d. None of the above

69. Which of the following shall not be considered as dead ends in drainage piping:

    a. Cleanout extensions

    b. Approved future fixture rough-ins

    c. Both A and B

    d. Neither A or B

70. Sewage pumps and sewage ejectors must automatically discharge the contents of their sumps into:

    a. A French drain
                             b. A holding tank

    c. A septic tank
                             d. The building drainage system

You've just completed the multiple-choice exam pertaining to sanitary drainage systems. How do you think you did? Before you check your answers, there is still a multiple choice and a true-false test to be taken. Proceed to the next quiz and answer the questions. When you complete them, you can proceed to the back of the chapter and check your answers.

FILL IN THE BLANK EXAM

1.   Building drains and sewers use the same _____ in determining the proper pipe size.

2.   _____ branches are the pipes branching off from a stack to accept the discharge from fixture drains.

3.   The number of fixture units allowed on a horizontal branch is determined by _____ and _____.

4.   A typical grade for drainage pipe is _____ inch of fall per foot.

5.   A drainage pipe installed underground must have a minimum diameter of _____ inches.

6.   All drainage piping must be protected from the effects of _____.

7.   Both horizontal and _____ pipes require support.

8.   Copper pipe should be hung with _____ hangers.

9.   You may not _____ the size of a drainage pipe as it heads for the waste-disposal site.

10.   When installed vertically, _____ must be supported every 15 feet.

Cross off the answers as you use them

pipe size	.25	criteria	pitch	copper	flooding
Horizontal	vertical	cast iron	reduce	two	

TRUE-FALSE EXAM

1. It is a code violation to install galvanized wrought iron or galvanized steel pipe below grade.
    True   False

2. The use of ABS or PVC pipe is limited to structures that do not have more than two stories above ground.
    True   False

3. Fittings used on screwed pipe must be of the recessed drainage type.
    True   False

4. A fixture trap with a diameter of 1 1/4 inches may handle up to two fixture units.
    True   False

5. A fixture trap with a diameter of 1 1/2 inches may handle up to three fixture units.
    True   False

6. A fixture trap with a diameter of 2 inches may handle up to six fixture units.
    True   False

7. A fixture trap with a diameter of 3 inches may handle up to six fixture units.
    True   False

8. A fixture trap with a diameter of 4 inches may handle up to eight fixture units.
    True   False

9. A bidet has a fixture-unit rating of 2 and requires a 1 1/2-inch trap.
    True   False

10. An automatic clothes washer requires a 2-inch trap.
    True   False

11. Trap size shall be consistent with the fixture outlet size.
    True   False

12. Knowing the amount of fall placed on the pipe is a component of sizing pipe.
    True   False

13. The minimum size of any building drain serving a water closet shall be 4 inches.
    True   False

14. When sizing a stack, you can base your decision on just the total number of fixture units carried by the stack.

    True   False

15. Code requirements are always the same in every jurisdiction.

    True   False

16. A typical grade for drainage pipe is .25 inch of fall per foot.

    True   False

17. Multiple buildings on the same building lot may share a common building sewer connecting to a public sewer.

    True   False

18. When mechanical joints are used, they must be made with an approved elastomeric seal.

    True   False

19. A long-sweep fitting can be used to change the direction of a horizontal pipe.

    True   False

20. You may use a double sanitary tee in a back-to-back situation if the fixtures being served are of a blowout or pump type.

    True   False

21. Sizing pipe is extremely difficult.

    True   False

22. One determination in sizing pipes is the drainage load.

    True   False

23. In a residential dwelling, the toilet has a fixture-unit value of three.

    True   False

24. A normal grade is generally .25 inch to the foot, but the fall can be steeper or shallower.

    True   False

25. Vertical branches are the pipes branching off from a stack to accept the discharge from fixture drains.

    True   False

MULTIPLE-CHOICE ANSWERS

1. c	15. b	29. d	43. b	57. b
2. a	16. c	30. c	44. a	58. a
3. c	17. b	31. b	45. b	59. d
4. d	18. a	32. c	46. b	60. d
5. a	19. b	33. c	47. a	61. d
6. b	20. a	34. b	48. a	62. d
7. c	21. b	35. a	49. d	63. d
8. d	22. b	36. b	50. d	64. a
9. b	23. b	37. a	51. c	65. d
10. c	24. b	38. b	52. d	66. d
11. c	25. b	39. a	53. b	67. a
12. b	26. a	40. c	54. c	68. c
13. a	27. b	41. c	55. d	69. c
14. c	28. b	42. b	56. b	70. d

FILL IN THE BLANK ANSWERS

1. criteria
2. Horizontal
3. pipe size and pitch
4. .25
5. 2
6. flooding
7. vertical
8. copper
9. reduce
10. cast iron

TRUE-FALSE ANSWERS

1. True	8. False	14. False	20. False
2. False	9. False	15. False	21. False
3. True	10. True	16. True	22. True
4. False	11. True	17. False	23. False
5. False	12. True	18. True	24. True
6. False	13. False	19. True	25. False
7. False			

TABLE 7.1 Drainage fixture units for fixtures and groups. *Copyright 2006, International Code Council, Inc., Falls Church, Virginia. Reproduced with permission. All rights reserved.*

FIXTURE TYPE	DRAINAGE FIXTURE UNIT VALUE AS LOAD FACTORS	MINIMUM SIZE OF TRAP (inches)
Automatic clothes washers, commercial[a,g]	3	2
Automatic clothes washers, residential[g]	2	2
Bathroom group as defined in Section 202 (1.6 gpf water closet)[f]	5	—
Bathroom group as defined in Section 202 (water closet flushing greater than 1.6 gpf)[f]	6	—
Bathtub[b] (with or without overhead shower or whirpool attachments)	2	$1\frac{1}{2}$
Bidet	1	$1\frac{1}{4}$
Combination sink and tray	2	$1\frac{1}{2}$
Dental lavatory	1	$1\frac{1}{4}$
Dental unit or cuspidor	1	$1\frac{1}{4}$
Dishwashing machine,[c] domestic	2	$1\frac{1}{2}$
Drinking fountain	$\frac{1}{2}$	$1\frac{1}{4}$
Emergency floor drain	0	2
Floor drains	2	2
Kitchen sink, domestic	2	$1\frac{1}{2}$
Kitchen sink, domestic with food waste grinder and/or dishwasher	2	$1\frac{1}{2}$
Laundry tray (1 or 2 compartments)	2	$1\frac{1}{2}$

Fixture type	Drainage fixture unit value	Minimum size of trap (inches)
Lavatory	1	$1^1/_4$
Shower	2	$1^1/_2$
Service sink	2	$1^1/_2$
Sink	2	$1^1/_2$
Urinal	4	Note d
Urinal, 1 gallon per flush or less	2[e]	Note d
Urinal, nonwater supplied	0.5	Note d
Wash sink (circular or multiple) each set of faucets	2	$1^1/_2$
Water closet, flushometer tank, public or private	4[e]	Note d
Water closet, private (1.6 gpf)	3[e]	Note d
Water closet, private (flushing greater than 1.6 gpf)	4[e]	Note d
Water closet, public (1.6 gpf)	4[e]	Note d
Water closet, public (flushing greater than 1.6 gpf)	6[e]	Note d

For SI: 1 inch = 25.4 mm, 1 gallon = 3.785 L (gpf = gallon per flushing cycle).

a. For traps larger than 3 inches, use Table 709.2.

b. A showerhead over a bathtub or whirlpool bathtub attachment does not increase the drainage fixture unit value.

c. See Sections 709.2 through 709.4 for methods of computing unit value of fixtures not listed in this table or for rating of devices with intermittent flows.

d. Trap size shall be consistent with the fixture outlet size.

e. For the purpose of computing loads on building drains and sewers, water closets and urinals shall not be rated at a lower drainage fixture unit unless the lower values are confirmed by testing.

f. For fixtures added to a dwelling unit bathroom group, add the dfu value of those additional fixtures to the bathroom group fixture count.

g. See Section 406.3 for sizing requirements for fixture drain, branch drain, and drainage stack for an automatic clothes washer standpipe.

TABLE 7.2 Drainage fixture units for fixture drains or traps.
Copyright 2006, International Code Council, Inc., Falls Church, Virginia.
Reproduced with permission. All rights reserved.

FIXTURE DRAIN OR TRAP SIZE (inches)	DRAINAGE FIXTURE UNIT VALUE
$1^1/_4$	1
$1^1/_2$	2
2	3
$2^1/_2$	4
3	5
4	6

For SI: 1 inch = 25.4 mm.

TABLE 7.3 Minimum capacity of sewage pump or sewage
ejector. *Copyright 2006, International Code Council, Inc., Falls Church,*
Virginia. Reproduced with permission. All rights reserved.

DIAMETER OF THE DISCHARGE PIPE (Inches)	CAPACITY OF PUMP OR EJECTOR (gpm)
2	21
$2^1/_2$	30
3	46

For SI: 1 inch = 25.4 mm, 1 gallon per minute = 3.785 L/m.

TABLE 7.4 Horizontal fixture branches and stacks. *Copyright 2006, International Code Council, Inc., Falls Church, Virginia. Reproduced with permission. All rights reserved.*

DIAMETER OF PIPE (inches)	Total for horizontal branch	MAXIMUM NUMBER OF DRAINAGE FIXTURE UNITS (dfu)			
		Total discharge into one branch interval	Stacks[b]		
			Total for stack of three branch intervals or less	Total for stack greater than three branch intervals
$1\frac{1}{2}$	3	2	4	8
2	6	6	10	24
$2\frac{1}{2}$	12	9	20	42
3	20	20	48	72
4	160	90	240	500
5	360	200	540	1,100
6	620	350	960	1,900
8	1,400	600	2,200	3,600
10	2,500	1,000	3,800	5,600
12	2,900	1,500	6,000	8,400
15	7,000	Note c	Note c	Note c

For SI: 1 inch = 25.4 mm.

a. Does not include branches of the building drain. Refer to Table 710.1(1).

b. Stacks shall be sized based on the total accumulated connected load at each story or branch interval. As the total accumulated connected load decreases, stacks are permitted to be reduced in size. Stack diameters shall not be reduced to less than one-half of the diameter of the largest stack size required.

c. Sizing load based on design criteria.

TABLE 7.5 Building drains and sewers. *Copyright 2006, International Code Council, Inc., Falls Church, Virginia. Reproduced with permission. All rights reserved.*

DIAMETER OF PIPE (inches)	MAXIMUM NUMBER OF DRAINAGE FIXTURE UNITS CONNECTED TO ANY PORTION OF THE BUILDING DRAIN OR THE BUILDING SEWER, INCLUDING BRANCHES OF THE BUILDING DRAIN[a]			
	Slope per foot			
	$^1/_{16}$ inch	$^1/_8$ inch	$^1/_4$ inch	$^1/_2$ inch
$1^1/_4$	—	—	1	1
$1^1/_2$	—	—	3	3
2	—	—	21	26
$2^1/_2$	—	—	24	31
3	—	36	42	50
4	—	180	216	250
5	—	390	480	575
6	—	700	840	1,000
8	1,400	1,600	1,920	2,300
10	2,500	2,900	3,500	4,200
12	3,900	4,600	5,600	6,700
15	7,000	8,300	10,000	12,000

For SI: 1 inch = 25.4 mm, 1 inch per foot = 83.3 mm/m.

a. The minimum size of any building drain serving a water closet shall be 3 inches.

TABLE 7.6 Stack sizes for bedpan steamers and boiling-type sterilizers. (Number of connections of various sizes permitted to various-sized sterilizer vent stacks). *Copyright 2006, International Code Council, Inc., Falls Church, Virginia. Reproduced with permission. All rights reserved.*

STACK SIZE (inches)	CONNECTION SIZE		
	1½"		2"
1½[a]	1	or	0
2[a]	2	or	1
2[b]	1	and	1
3[a]	4	or	2
3[b]	2	and	2
4[a]	8	or	4
4[b]	4	and	4

For SI: 1 inch = 25.4 mm.
a. Total of each size.
b. Combination of sizes.

TABLE 7.7 Stack sizes for pressure sterilizers. (Number of connections of various sizes permitted to various-sized sterilizer vent stacks). *Copyright 2006, International Code Council, Inc., Falls Church, Virginia. Reproduced with permission. All rights reserved.*

STACK SIZE (inches)	CONNECTION SIZE			
	¾"	1"	1¼"	1½"
1½[a]	3 or	2 or	1	—
1½[b]	2 and	1	—	—
2[a]	6 or	3 or	2 or	1
2[b]	3 and	2	—	—
2[b]	2 and	1 and	1	—
2[b]	1 and	1 and	—	1
3[a]	15 or	7 or	5 or	3
3[b]	1 and	1 and / 5 and	2 and —	2 / 1

For SI: 1 inch = 25.4 mm.
a. Total of each size.
b. Combination of sizes.

TABLE 7.8   Pipe fittings. *Copyright 2006, International Code Council, Inc., Falls Church, Virginia. Reproduced with permission. All rights reserved.*

MATERIAL	STANDARD
Acrylonitrile butadiene styrene (ABS) plastic pipe	ASTM D 2661; ASTM D 3311; CSA B181.1
Cast iron	ASME B 16.4; ASME B 16.12; ASTM A 74; ASTM A 888; CISPI 301
Coextruded composite ABS DWV schedule 40 IPS pipe (solid or cellular core)	ASTM D 2661; ASTM D 3311; ASTM F 628
Coextruded composite PVC DWV schedule 40 IPS-DR, PS140, PS200 (solid or cellular core)	ASTM D 2665; ASTM D 3311; ASTM F 891
Coextruded composite ABS sewer and drain DR-PS in PS35, PS50, PS100, PS140, PS200	ASTM D 2751
Coextruded composite PVC sewer and drain DR-PS in PS35, PS50, PS100, PS140, PS200	ASTM D 3034
Copper or copper alloy	ASME B 16.15; ASME B 16.18; ASME B 16.22; ASME B 16.23; ASME B 16.26; ASME B 16.29
Glass	ASTM C 1053
Gray iron and ductile iron	AWWA C 110
Malleable iron	ASME B 16.3
Polyolefin	ASTM F 1412; CSA B181.3
Polyvinyl chloride (PVC) plastic	ASTM D 2665; ASTM D 3311; ASTM F 1866
Stainless steel drainage systems, Types 304 and 316L	ASME A 112.3.1
Steel	ASME B 16.9; ASME B16.11; ASME B16.28

TABLE 7.9 Fittings for change in direction. *Copyright 2006, International Code Council, Inc., Falls Church, Virginia. Reproduced with permission. All rights reserved.*

TYPE OF FITTING PATTERN	CHANGE IN DIRECTION		
	Horizontal to vertical	Vertical to horizontal	Horizontal to horizontal
Sixteenth bend	X	X	X
Eighth bend	X	X	X
Sixth bend	X	X	X
Quarter bend	X	X[a]	X[a]
Short sweep	X	X[a,b]	X[a]
Long sweep	X	X	X
Sanitary tee	X[c]	—	—
Wye	X	X	X
Combination wye and eighth bend	X	X	X

For SI: 1 inch = 25.4 mm.

a. The fittings shall only be permitted for a 2-inch or smaller fixture drain.

b. Three inches or larger.

c. For a limitation on double sanitary tees, see Section 706.3.

TABLE 7.10 Above-ground drainage and vent pipe choices.
Copyright 2006, International Code Council, Inc., Falls Church, Virginia.
Reproduced with permission. All rights reserved.

MATERIAL	STANDARD
Acrylonitrile butadiene styrene (ABS) plastic pipe	ASTM D 2661; ASTM F 628; CSA B181.1
Brass pipe	ASTM B 43
Cast-iron pipe	ASTM A 74; ASTM A 888; CISPI 301
Coextruded composite ABS DWV schedule 40 IPS pipe (solid)	ASTM F 1488
Coextruded composite ABS DWV schedule 40 IPS pipe (cellular core)	ASTM F 1488
Coextruded composite PVC DWV schedule 40 IPS pipe (solid)	ASTM F 1488
Coextruded composite PVC DWV schedule 40 IPS pipe (cellular core)	ASTM F 891; ASTM F 1488
Coextruded composite PVC IPS-DR, PS140, PS200 DWV	ASTM F 1488
Copper or copper-alloy pipe	ASTM B 42; ASTM B 302
Copper or copper-alloy tubing (Type K, L, M or DWV)	ASTM B 75; ASTM B 88; ASTM B 251; ASTM B 306
Galvanized steel pipe	ASTM A 53
Glass pipe	ASTM C 1053
Polyolefin pipe	CSA B181.3
Polyvinyl chloride (PVC) plastic pipe (Type DWV)	ASTM D 2665; ASTM D 2949; ASTM F 1488; CSA B181.2
Stainless steel drainage systems, Types 304 and 316L	ASME A112.3.1

TABLE 7.11   Underground building drainage and vent pipe.
Copyright 2006, International Code Council, Inc., Falls Church, Virginia.
Reproduced with permission. All rights reserved.

MATERIAL	STANDARD
Acrylonitrile butadiene styrene (ABS) plastic pipe	ASTM D 2661; ASTM F 628; CSA B181.1
Asbestos-cement pipe	ASTM C 428
Cast-iron pipe	ASTM A 74; ASTM A 888; CISPI 301
Coextruded composite ABS DWV schedule 40 IPS pipe (solid)	ASTM F 1488
Coextruded composite ABS DWV schedule 40 IPS pipe (cellular core)	ASTM F 1488
Coextruded composite PVC DWV schedule 40 IPS pipe (solid)	ASTM F 1488
Coextruded composite PVC DWV schedule 40 IPS pipe (cellular core)	ASTM F 891; ASTM F 1488
Coextruded composite PVC IPS-DR, PS140, PS200 DWV	ASTM F 1488
Copper or copper-alloy tubing (Type K, L, M or DWV)	ASTM B 75; ASTM B 88; ASTM B 251; ASTM B 306
Polyolefin pipe	ASTM F 1412; CSA B181.3
Polyvinyl chloride (PVC) plastic pipe (Type DWV)	ASTM D 2665; ASTM D 2949; CSA B181.2
Stainless steel drainage systems, Type 316L	ASME A112.3.1

TABLE 7.12 Choices of piping for building sewers.
Copyright 2006, International Code Council, Inc., Falls Church,
Virginia. Reproduced with permission. All rights reserved.

MATERIAL	STANDARD
Acrylonitrile butadiene styrene (ABS) plastic pipe	ASTM D 2661; ASTM D 2751; ASTM F 628
Asbestos-cement pipe	ASTM C 428
Cast-iron pipe	ASTM A 74; ASTM A 888; CISPI 301
Coextruded composite ABS DWV schedule 40 IPS pipe (solid)	ASTM F 1488
Coextruded composite ABS DWV schedule 40 IPS pipe (cellular core)	ASTM F 1488
Coextruded composite PVC DWV schedule 40 IPS pipe (solid)	ASTM F 1488
Coextruded composite PVC DWV schedule 40 IPS pipe (cellular core)	ASTM F 891; ASTM F 1488
Coextruded composite PVC IPS-DR, PS140, PS200, DWV	ASTM F 1488
Coextruded composite ABS sewer and drain DR-PS in PS35, PS50, PS100, PS140, PS200	ASTM F 1488
Coextruded composite PVC sewer and drain DR-PS in PS35, PS50, PS100, PS140, PS200	ASTM F 1488
Coextruded PVC sewer and drain PS25, PS50, PS100 (cellular core)	ASTM F 891
Concrete pipe	ASTM C14; ASTM C76; CSA A257.1M; CSA A257.2M
Copper or copper-alloy tubing (Type K or L)	ASTM B 75; ASTM B 88; ASTM B 251
Polyethylene (PE) plastic pipe (SDR-PR)	ASTM F 714
Polyvinyl chloride (PVC) plastic pipe (Type DWV, SDR26, SDR35, SDR41, PS50 or PS100)	ASTM D 2665; ASTM D 2949; ASTM D 3034; CSA B182.2; CSA B182.4
Stainless steel drainage systems, Types 304 and 316L	ASME A112.3.1
Vitrified clay pipe	ASTM C 4; ASTM C 700

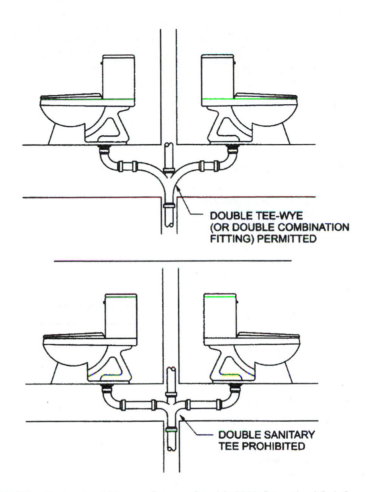

FIGURE 7.1   Back-to-back blowout fixtures. *Copyright 2002, International Code Council, Inc., Falls Church, Virginia. Reproduced with permission. All rights reserved.*

—NOTES—

Chapter 8
INDIRECT AND SPECIAL WASTES

MULTIPLE-CHOICE EXAM

1. An indirect-waste receptor must be trapped if the drain from certain fixtures is more than _____ feet long.

    a. 4   b. 5

    c. 3   d. 2

2. The safest method of indirect waste disposal is accomplished by using:

    a. an air gap   b. a sink trap

    c. a sink drain   d. a relief valve

3. Which one of these wastes may have a harmful effect on plumbing or waste-disposal system?

    a. water waste   b. special waste

    c. sewer waste   d. indirect waste

4. Indirect-waste requirements pertain to which of the following:

    a. Refrigeration   b. toilet facilities

    c. Clothes washer drain   d. None of the above

5. If a floor drain is located within an area subject to freezing, the waste line serving the drain must not be:

    a. Over 3 inches   b. an air-gap

    c. an air-break   d. trapped

6. The standpipe for an automatic clothes washer must have a minimum diameter of _____ inches.

    a. 6   b. 3

    c. 2   d. 4

7. _____ water waste from a potable source must be piped to indirect waste through an air gap.

    a. Clean   b. Dirty

    c. Potable   d. Dingy

8. The direct connection of any _____ to the sanitary drainage system is likely to be prohibited.

    a. toilet   b. dishwasher

    c. sink   d. water filters

9.   An approved fixture or device that is used to accept the discharge from indirect waste pipes is what:

    a.   Air-gap   b.   Relief valve

    c.   Air-break   d.   A receptor

10.   Code generally requires all receptors to be equipped with a means of preventing solids with diameters of _____ inch or larger from entering the drainage system.

    a.   .5   b.   025

    c.   75   d.   1.4

TRUE-FALSE EXAM

1.   The risk of an air break is the possibility of a backup.
      True   False

2.   A tailpiece is an approved receptor of a kitchen sink.
      True   False

3.   If a clear-water waste receptor is located in a floor, some codes require the lip of the receptor to extend at least 2.5 inches above the floor.
      True   False

4.   It is not acceptable for a garbage disposal to receive waste that has passed through an air gap.
      True   False

5.   Domestic dishwashing machines must discharge indirectly through an air gap or air break.
      True   False

6.   If you are concerned with sizing a particular waste receptor, you should ask the homeowner what size he thinks you should use.
      True   False

7.   Codes do not prohibit the installation of an indirect-waste receptor in any room containing toilet facilities.
      True   False

8.   Special wastes may have a harmful effect on a plumbing or waste-disposal system.
      True   False

9.   Buildings that are considered to have a need for special-wastes plumbing are often required to have two plumbing systems.

     True   False

10. Special wastes must be neutralized, diluted, or otherwise treated before it can enter a sanitary drainage system.

     True   False

11. Domestic dishwashing machines must discharge directly through an air gap or air break into a standpipe.

     True   False

12. Some code regions require that a discharge pipe terminate at least two inches above the receptor.

     True   False

13. A photographic lab is a possible location for special waste piping.

     True   False

14. A swimming pool is an example of where clear water waste would come from.

     True   False

15. An air gap and an air break are interchangeable and have the same meaning.

     True   False

MULTIPLE-CHOICE ANSWERS

1. d	3. b	5. d	7. a	9. d
2. a	4. c	6. c	8. b	10. a

TRUE-FALSE ANSWERS

1. True	5. True	9. True	13. True
2. True	6. False	10. True	14. True
3. False	7. False	11. False	15. False
4. False	8. True	12. True	

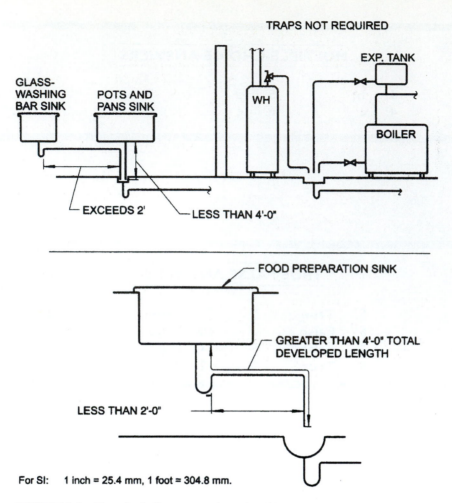

FIGURE 8.1   Traps for indirect waste pipes. *Copyright 2002, International Code Council, Inc., Falls Church, Virginia. Reproduced with permission. All rights reserved.*

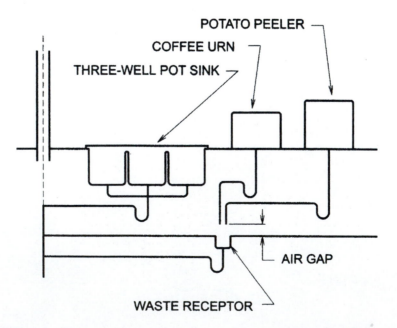

FIGURE 8.2   Restaurant kitchen with indirect waste and an air gap. *Copyright 2002, International Code Council, Inc., Falls Church, Virginia. Reproduced with permission. All rights reserved.*

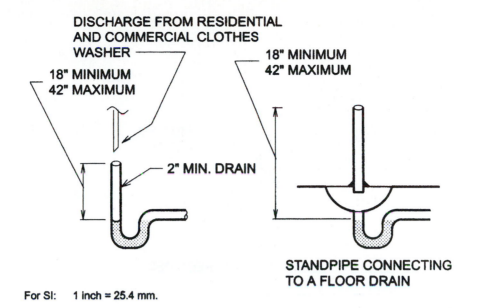

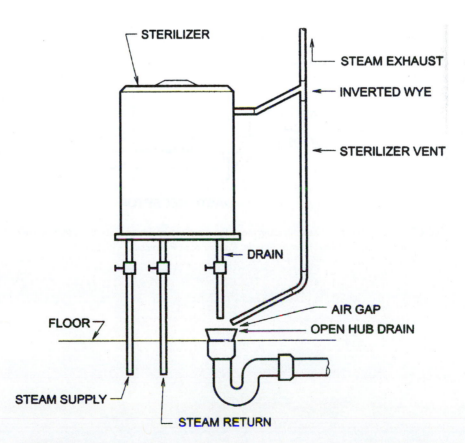

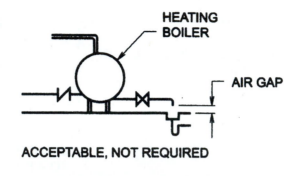

ACCEPTABLE, NOT REQUIRED

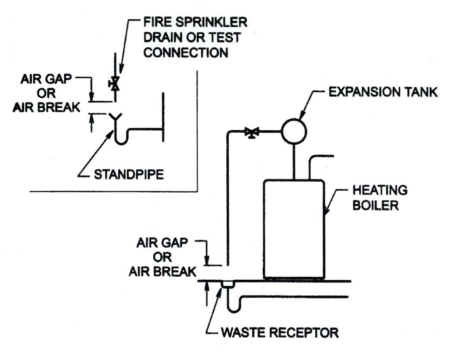

FIGURE 8.5 Indirect waste arrangement for clear-water waste. *Copyright 2002, International Code Council, Inc., Falls Church, Virginia. Reproduced with permission. All rights reserved.*

Chapter 9
VENTS

The installation of vents can be complicated. Without proper venting, a drainage system will not work very well. Even if you didn't have to worry about plumbing inspectors rejecting your work, you would certainly have dissatisfied customers should you fail to vent plumbing fixtures. Many people, licensed plumbers included, don't understand fully the primary purpose of vents. Do you know the main reason why vents are installed for each plumbing fixture?

If you asked 100 people why plumbing fixtures are vented, a large percentage of them would most likely agree that vents are installed to keep smells out of buildings. Many people think that vents are installed to carry sewer gas from the sewer to the outside air. They think that without the vents, sewer gas would enter the building and become a problem. Would you agree with this group of people?

Vents in a plumbing system are installed to protect the seals of traps. It is the trap of a fixture that prevents sewer gas from entering a building. While a majority of people thinks of traps as being installed only to prevent foreign objects from entering the drainage pipes, the traps are actually there to keep odors and dangerous gases from entering habitable space. However, without the proper venting, a trap seal could be siphoned down the drain, leaving the building unprotected from sewer gas and odors.

Vents provide air circulation to a drainage system. They allow the drains to flow faster, and they prevent siphonic action in the traps. These are the two main purposes of vents in a plumbing system. Fixtures that are not vented adequately drain slowly. This same principle can be demonstrated with a can of juice.

Have you ever punched a hole in the top of a can of juice or motor oil and tried to pour the contents from the can? If you have, you know that the liquid comes out slowly and in a jerky motion. But if you punch a second hole in the top of the can opposite the location of the first hole, the liquid pours out faster and more smoothly. This is a fine example of how a plumbing vent improves the flow of drainage. The second hole made in the top of the can acts as a vent for the hole used to pour from. The same basic principle applies to a plumbing fixture. Without a vent, a fixture can't drain as well. The key is air circulation.

Not all plumbing vents are vertical pipes extending to open air. Some are wet vents, meaning that the vent for one fixture serves as a drain for another fixture. In combination-waste-and-vent systems, the drains are oversized to allow air circulation above the flow of drainage. There are many options for venting a plumbing system, and if you want to pass your licensing exam, you had better know most of them.

Let's discuss the issue of protecting trap seals. When a trap is not properly vented, there is a risk that the water that forms the seal will be sucked down the drain, leaving the trap without enough water to block odors and gases. Most people have siphoned some type of liquid at some time in their lives. As you probably know, once a siphon hose has been started, the flow in the hose will continue until air is introduced into the hose or until the hose is raised above the level of the liquid being siphoned. This same principle applies to a drainage system.

Plumbing drains are typically installed at levels lower than the fixtures they serve unless a pump is used to empty the fixture or sump. This means that the drain will not be raised above the level of the fixture, so siphonic action can't be broken in that way.

If you were to fill a bathtub with water and then open the drain, the water would pass through the trap and down the drain. Without a vent, the volume of water draining down the pipe could create a siphonic action and pull all water from above the drain into the drainage system, including the water in the trap. Should this happen, the fixture would be left unprotected from odors and gases.

The trap might be replenished when a small quantity of water was run through it, but the next time a large volume of water was drained, the trap could be emptied again. This, of course, is not an acceptable risk. Therefore, the trap should be vented to avoid the problem. The vent allows air circulation in the drainage system, and the air prevents the possibility of siphonic action in the drain.

Now you know more about the true purpose of vents than some master plumbers do. I have known many plumbers who didn't understand why vents were used. They installed them because code requirements dictated the need for them, but the plumbers couldn't explain why the vents were needed. You are about to be tested on the proper installation of vents, and I thought the knowledge might be easier to gain if you had a good understanding of why vents are needed and how they work. Now, let's get on with the multiple-choice exam.

MULTIPLE-CHOICE EXAM

1. Which of the following types of vent systems must be independent of a sanitary vent system:
   a. Wet-vent system
   b. Dry-vent system
   c. Chemical-vent system
   d. None of the above

2. A vent that connects to one or more individual vents with a stack vent or a vent stack is known as:
   a. A yoke vent
   b. A wet vent
   c. A main vent
   d. A branch vent

3. The primary vent in a venting system to which various vent branches are connected is called a:
   a. Main vent
   b. Stack vent
   c. Vent stack
   d. Yoke vent

4. A vent that serves only one fixture drain is called:
   a. A fixture branch
   b. A circuit vent
   c. An individual vent
   d. A common vent

5. When a soil or waste stack is extended above the highest horizontal drain connected to the stack, which of the following is created:
   a. A code violation
   b. A vent stack
   c. A stack vent
   d. A relief vent

6. Vents that connect to horizontal drainage branches and vent more than one trap are called:
   a. Fixture branches
   b. Circuit vents
   c. Vent stacks
   d. Relief vents

7. Vents are intended to protect:
   a. Traps
   b. Trap seals
   c. Fixture outlets
   d. All of the above

8. Vents used to protect against pressure changes in a stack are called:

    a. Pneumatic vents   b. Wet vents

    c. Circuit vents   d. Yoke vents

9. A vertical pipe installed to provide air circulation to and from the drainage system is called:

    a. A stack vent   b. A vent stack

    c. A leader   d. A conductor

10. An auxiliary vent that provides additional circulation of air in or between a drainage and vent system is called:

    a. A branch vent   b. An island vent

    c. A relief vent   d. A common vent

11. The section of pipe in a drainage system that runs from the trap of a fixture to a junction with any other drain pipe is known as:

    a. A horizontal branch   b. A fixture drain

    c. A fixture branch   d. None of the above

12. Drains that serve one or more fixtures that discharge into another drain or into a stack are called:

    a. Wet vents   b. Fixture drains

    c. Fixture branches   d. None of the above

13. Vents that serve the containers of sewage ejectors or similar equipment and terminate separately to open air are called:

    a. Yoke vents   b. Pit vents

    c. Sump vents   d. Pump vents

14. The minimum size of any vent pipe shall not be less than:

    a. 1 inch   b. 1 1/2 inches

    c. 1 1/2 inches   d. 2 inches

15. The size of any vent with a developed length of more than 40 feet must be increased by:

    a. 1 inch   b. 1 pipe size

    c. 50 percent   d. None of the above

16. When determining the developed length of a vent, one must measure from the furthest point of the vent connection to the drainage system to the point of:

    a. Connection to a vent stack   b. Connection to a stack vent

    c. Termination into open air   d. Any of the above

17. The diameter of vents must be, at a minimum, equal to:
    a. The size of the drain being served
    b. One-half the size of the drain being served
    c. Twice the size of the drain being served
    d. None of the above

18. Individual vents must be connected to the _____ of the trap or trapped fixture being served. Fill in the blank:
    a. Top
    b. Bottom
    c. Fixture drain
    d. Fixture branch

19. A vent system for chemical waste must:
    a. Not be run horizontally
    b. Connect to a vent stack
    c. Connect to a stack vent
    d. Terminate separately through the roof to open air

20. A stack vent is the same as:
    a. A vent stack
    b. An individual vent
    c. A branch vent
    d. None of the above

21. The length of a pipeline that is measured along the centerline of the pipe and fittings is the:
    a. Combined length
    b. Determined length
    c. Developed length
    d. Maximum length

22. Vents must protect trap seals in such a way that the seal is not subjected to a pneumatic pressure differential equal to or exceeding a:
    a. 1-inch water column
    b. 1 1/2-inch water column
    c. 2-inch water column
    d. None of the above

23. All sanitary drainage systems that receive the discharge of a water closet are required to have:
    a. A vent stack
    b. A stack vent
    c. Either A or B
    d. Both A and B

24. Every vent stack must terminate:
    a. Through a roof
    b. Outdoors
    c. Near a window
    d. Both A and B

25.   When a drainage stack consists of five branch intervals or more, a _____ is required. Fill in the blank:

a.   Vent stack   b.   Stack vent

c.   Combined vent   d.   Relief vent

26.   All vent stacks must connect to:

a.   The base of a drainage stack   b.   The base of a sewer

c.   The base of a fixture   d.   Any of the above

27.   When a vent stack is connected to a building drain, the connection must be made within _____ pipe diameters downstream of the drainage stack:

a.   2   b.   4

c.   5   d.   10

28.   If there are five or more branch intervals located above a horizontal offset of a drainage stack, the offsets must be:

a.   No more than 22 1/2 degrees

b.   Trapped

c.   Vented

d.   Eliminated

29.   The length of a vent terminal when it penetrates a roof must be no less than _____ inches above the roof. Fill in the blank:

a.   4   b.   6

c.   8   d.   12 or in compliance with local code requirements

30.   When a roof is used for some purpose other than only weather protection, the extension of a vent penetrating the roof must not be less than:

a.   6 inches   b.   12 inches

c.   4 feet   d.   7 feet

31.   Vent terminals must not be installed directly beneath any:

a.   Window   b.   Door

c.   Ventilating opening   d.   All of the above

32.   When a vent pipe is installed on the outside of a building where freezing temperatures may occur, the vent must be protected from freezing by which of the following means:

a.   Insulation   b.   Heat

c.   Either A or B   d.   Both A and B

33. Vent terminals may not be installed:

    a. In attics   b. Through walls

    c. Through roofs   d. All of the above

34. A vent may not terminate within _____ feet of a window unless it is extended to a point at least _____ feet above the window. Fill in the blanks:

    a. 5 and 2   b. 2 and 5

    c. 10 and 2   d. 10 and 5

35. Vents that terminate through a wall must be at least _____ feet from the lot line of the property. Fill in the blank:

    a. 5   b. 10

    c. 15   d. 20

36. Vents that terminate through a wall, must be at least _____ feet above the ground. Fill in the blank

    a. 5   b. 10

    c. 15   d. 20

37. Vent terminals may not terminate under the overhang of the building when the overhang:

    a. Extends less than 4 inches

    b. Contains soffit vents

    c. Will not protect the opening of the vent

    d. All of the above

38. Side-wall vents must be protected to prevent birds and rodents from:

    a. Entering the vent opening   b. Blocking the vent opening

    c. Both A and B   d. Either A or B

39. With one exception, all individual, branch, and circuit vents must:

    a. Connect to a vent stack   b. Connect to a stack vent

    c. Terminate into open air   d. Any of the above

40. All dry vents must rise vertically to a point at least _____ inches above the flood-level rim of the highest trap or trapped fixture being vented. Fill in the blank:

    a. 4   b. 6

    c. 12   d. 42

41. All vents should be graded and connected in a manner that will allow them to:

a. Drain back to a soil or waste pipe by gravity

b. Maintain a grade of 1/8 inch to the foot

c. Be rodded [not clear] out from the roof

d. Maintain a grade of 1/2 inch to the foot

42. Which of the following types of vents must, when connecting to a horizontal drain, connect above the centerline of the horizontal drain pipe:

a. Dry vents   b. Wet vents

c. Combination vents   d. All of the above

43. Horizontal vent pipes that form circuit vents must be connected at a point at least _____ inches above the flood-level rim of the highest fixture being vented. Fill in the blank:

a. 4   b. 6

c. 8   d. 42

44. Horizontal vent pipes that form branch vents must be connected at a point at least _____ inches above the flood-level rim of the highest fixture being vented. Fill in the blank:

a. 4   b. 6

c. 8   d. 42

45. Horizontal vent pipes that form relief vents must be connected at a point at least _____ inches above the flood-level rim of the highest fixture being vented. Fill in the blank:

a. 4   b. 6

c. 8   d. 42

46. When a vent pipe is connected to a vent stack or a stack vent, the connection must be made at least _____ inches above the flood-level rim of the highest fixture being vented. Fill in the blank:

a. 4   b. 6

c. 8   d. 42

47. Vents that are roughed in for future fixtures must:

a. Connect to the vent system

b. Be labeled as a vent

c. Both A and B

d. None of the above

48. Any combination of fixtures within _____ bathroom groups located on the same floor level may be wet-vented. Fill in the blank:

   a. 2   b. 3

   c. 4   d. Any of the above

49. A dry-vent connection to a wet vent must be an individual vent or a common vent to which of the following:

   a. Bidet   b. Lavatory

   c. Bathtub   d. Any of the above

50. A dry-vent connection to a wet vent must be an individual vent or a common vent to which of the following:

   a. Shower   b. Lavatory

   c. Bathtub   d. Any of the above

51. A wet vent with a diameter of 1 1/2 inches may accept the discharge of how many fixture units:

   a. 1   b. 2

   c. 4   d. 6

52. A wet vent with a diameter of 2 inches may accept the discharge of how many fixture units:

   a. 1   b. 2

   c. 4   d. 6

53. A wet vent with a diameter of 3 inches may accept the discharge of how many fixture units:

   a. 3   b. 4

   c. 6   d. 12

54. Individual vents may be used as a common vent to vent _____ traps or trapped fixtures located on the same floor level. Fill in the blank: [what exception? Cover elsewhere?]

   a. 2   b. 3

   c. 4   d. 5

55. Fixtures being served by a common vent and connecting at the same level shall connect to the vent at the interconnection of:

   a. The traps   b. The fixture branches

   c. The fixture drains   d. The stack vent

56. A common vent with a diameter of 1 1/2 inches may accept the maximum discharge from upper fixture drains from how many fixture units:

   a.   1   b.   2

   c.   3   d.   4

57. A common vent with a diameter of 2 inches may accept the maximum discharge from upper fixture drains from how many fixture units:

   a.   2   b.   4

   c.   6   d.   8

58. A common vent with a diameter of 3 inches may accept the maximum discharge from upper fixture drains from how many fixture units:

   a.   2   b.   4

   c.   6   d.   8

59. What is the maximum number of fixtures that may be connected to a horizontal branch drain and be vented with a circuit vent:

   a.   3   b.   4

   c.   6   d.   8

60. When a circuit vent is used, each fixture drain shall connect _____ to the horizontal branch being circuit-vented. Fill in the blank:

   a.   Vertically   b.   Horizontally

   c.   Indirectly   d.   None of the above

61. Circuit vents are not allowed to:

   a.   Run for more than 10 feet

   b.   Receive the discharge from any soil pipe

   c.   Run horizontally

   d.   Connect to a vent stack

62. Circuit vents are not allowed to:

   a.   Run for more than 10 feet

   b.   Receive the discharge from any waste pipe

   c.   Run horizontally

   d.   Connect to a stack vent

63.   The maximum slope allowed on a vent section of a horizontal branch drain is:

    a.   2 percent of slope   b.   4 percent of slope

    c.   8 percent of slope   d.   12 percent of slope

64.   When a circuit-vented horizontal branch receives the discharge of _____ or more water closets, a relief vent must be provided. Fill in the blank:

    a.   2   b.   3

    c.   4   d.   6

65.   Crown venting is not allowed within _____ pipe diameters of the trap weir. Fill in the blank:

    a.   1 1/2   b.   2

    c.   2 1/2   d.   3

66.   Vents for fixture drains, excluding fixtures with integral traps, must connect above:

    a.   The flood-level rim of the fixture

    b.   The weir of the trap being vented

    c.   The fixture drain

    d.   None of the above

    To determine the distance of a fixture trap from its vent, a plumber will normally consult a table in a code book. These tables, however, may not always be available, such as on the day of a licensing exam. To determine the distance limitation between a trap and its fixture, you must know the size of the trap, the size of the fixture drain, and the slope of the pipe. With that information provided, you should be able to decide how far a trap can be placed from its vent. The next questions will test your knowledge about this issue. I will provide the information needed to make a correct decision, but you must know the maximum distances allowed by the plumbing code. Here is your first scenario.

67.   You have a fixture with a 1 1/2-inch trap and drain. The slope of the pipe is 1/4 inch per foot. How far from the vent may the trap for the fixture be placed:

    a.   3 1/2 feet   b.   4 feet

    c.   5 feet   d.   6 feet

68.   You have a fixture with a 1 1/4-inch trap and drain. The slope of the pipe is 1/4 inch per foot. How far from the vent may the trap for the fixture be placed:

    a.   3 1/2 feet   b.   4 feet

    c.   5 feet   d.   6 feet

69.   You have a fixture with a 1 1/4-inch trap and a 1 1/2-inch drain. The slope of the pipe is 1/4 inch per foot. How far from the vent may the trap for the fixture be placed:

      a.   3 1/2 feet   b.   4 feet

      c.   5 feet   d.   6 feet

70.   You have a fixture with a 1 1/2-inch trap and a 2-inch drain. The slope of the pipe is 1/4 inch per foot. How far from the vent may the trap for the fixture be placed:

      a.   42 inches   b.   4 feet

      c.   6 feet   d.   8 feet

71.   You have a fixture with a 2-inch trap and a 2-inch drain. The slope of the pipe is 1/4 inch per foot. How far from the vent may the trap for the fixture be placed:

      a.   42 inches   b.   4 feet

      c.   6 feet   d.   8 feet

72.   You have a fixture with a 3-inch trap and a 3-inch drain. The slope of the pipe is 1/8 inch per foot. How far from the vent may the trap for the fixture be placed:

      a.   6 feet   b.   8 feet

      c.   10 feet   d.   12 feet

73.   You have a fixture with a 4-inch trap and a 4-inch drain. The slope of the pipe is 1/8 inch per foot. How far from the vent may the trap for the fixture be placed:

      a.   6 feet   b.   8 feet

      c.   10 feet   d.   12 feet

74.   A relief vent must be provided for every drainage stack that has more than _____ branch intervals. Fill in the blank:

      a.   4   b.   5

      c.   8   d.   10

75.   When a relief vent is provided for a drainage stack, the relief vent must:

      a.   Connect to a stack vent   b.   Connect to a vent stack

      c.   Either A or B   d.   Be installed horizontally

! Codealert

Stack-type air admittance valves need to conform to ASSE 1050. Individual and branch-type air admittance valves must conform to ASSE 1051.

76. Which of the following fixtures may be served by a combination-waste-and-vent system:

    a. Floor drains   b. Sinks

    c. Lavatories   d. Any of the above

77. Which of the following fixtures may not be served by a combination-waste-and-vent system:

    a. Standpipes   b. Water closets

    c. Lavatories   d. A and B

78. The maximum vertical distance for a vertical pipe in a combination-waste-and-vent system is:

    a. 30 inches   b. 4 feet

    c. 8 feet   d. 10 feet

79. The maximum slope allowed on a horizontal combination-waste-and-vent pipe is:

    a. 2 percent   b. 3 percent

    c. 4 percent   d. 6 percent

80. The only vertical pipes allowed in a combination-waste-and-vent system are the connections between horizontal pipes and the fixture branches of:

    a. Lavatories   b. Sinks

    c. Standpipes   d. All of the above

81. When there is a danger of frost closure, as determined by the plumbing code, in a vent, the vent terminal must be a minimum of _____ inches in diameter, and if the vent size must be increased to accomplish this, the increase shall take place at least _____ inches below the roof or inside wall where the vent penetrates:

    a. 1 1/2 and 6   b. 2 and 12

    c. 3 and 12   d. 4 and 12

82. A waste stack vent must be installed:

    a. With offsets

    b. In conjunction with a chemical stack

    c. Vertically

    d. In buildings with more than two stories above ground

83. When a stack vent is provided for a waste stack, the stack vent must be sized to be:

    a. No more than 1 pipe diameter smaller than the waste stack

    b. The same size as the waste stack

    c. 1 pipe size larger than the waste stack

    d. No less than 4 inches in diameter

84. Island fixture venting is allowed for which of the following fixtures:

    a. Sinks   b. Bidets

    c. Floor drains   d. All of the above

85. Island fixture venting is allowed for which of the following fixtures:

    a. Sinks   b. Bidets

    c. Lavatories   d. Either A or C

86. Island fixture venting is allowed for which of the following fixtures:

    a. Residential kitchen sinks

    b. Residential kitchen sinks with dishwasher waste connections

    c. Residential kitchen sinks with garbage disposers

    d. Any of the above

87. Island vents must rise vertically to a point:

    a. Above the flood-level rim of the fixture being vented

    b. At least 6 inches above the flood-level rim of the fixture being vented

    c. Above the drainage outlet of the fixture being vented

    d. At least 12 inches above the base of the cabinet containing the fixture

88. When island vents are used, _____ must be installed and accessible:

    a. Cleanouts   b. Trap adapters

    c. Sanitary tees   d. None of the above

89. When sheet copper is used as a flashing material for vents, it must have a weight of not less than: [can't indent]

    a. 6 ounces per square foot   b. 8 ounces per square foot

    c. 10 ounces per square foot   d. 12 ounces per square foot

90. Sheet lead for vent-pipe flashings used in field-constructed flashings shall weigh not less than _____ pounds per square foot:

    a. 2   b. 3

    c. 5   d. 8

FILL IN THE BLANK EXAM

1. Vents play a vital role in the scheme of _____ plumbing.

2. In some jurisdictions, _____ waste and vent systems are used.

3. Antisiphon is another name or meaning for _____ .

4. When the primary vent is too far from the fixture, a _____ vent is needed.

5. _____ vents may tie into stack vents or vent stacks.

6. Branch vents are vents extending _____ and connecting multiple vents together.

7. Vent stacks run _____ and are sized a little differently.

8. The lower portion of a stack vent is called _____.

9. Vents protect trap seals, by regulating the _____ pressure applied to the seals.

10. _____ are allowed for sinks and lavatories.

Cross off the answers as you use them

drum traps	Circuit	horizontally	island vents	sanitary
combination	relief	vertically	a soil pipe	atmospheric

This concludes the multiple-choice and fill in the blank testing on vents. Depending on where you will be working and taking your licensing exam, there may be some differences between your local code and the questions provided in this exam. This same situation may apply to other exams in this book. Always confirm your answers with a current, local code reference.

We will now move onto the true-false test

TRUE-FALSE EXAM

1.   Every plumbing fixture, except as otherwise provided in the code, must be vented.

        True   False

2.   The use of ABS and PVC piping is limited to buildings 3 or fewer stories above ground.

        True   False

3.   A vent may exceed one-third the maximum horizontal length allowed by the code if the size of the vent is increased by one pipe size for its entire developed length.

        True   False

4.   All vent pipes must have a minimum slope of 1/4 inch per foot.

        True   False

5.   Every vent must rise a minimum of 42 inches above the flood-level rim of the fixture being served by the vent.

        True   False

6.   No vent may terminate above a roof with an exposed distance of less than 12 inches or as required by local codes.

        True   False

7.   All vents that terminate through a roof must extend 6 inches above the roof.

        True   False

8.   Any vent terminating above a roof must be at least 12 inches away from any vertical surface.

        True   False

9.   A vent installed within 10 feet of a roof used for purposes other than weather protection must extend at least 7 feet above the roof.

        True   False

10.   The minimum diameter for a vent that terminates above a roof in an area where frost closure may be a problem is 3 inches.

        True   False

11.   When a vent is being installed in an area that may be affected by frost closure, the vent must extend at least 24 inches above the roof.

        True   False

12. Drainage systems with 10 or more branch intervals must be installed in conjunction with a relief vent.

    True   False

13. Island vents must extend to the drainboard of the fixture being served.

    True   False

14. Combination-waste-and-vent systems are prohibited.

    True   False

15. Vent pipes may not be used to support flagpoles or similar items.

    True   False

16. Sewer gas can create an explosion when it is concentrated in a poorly ventilated area.

    True   False

17. Vents help fixtures drain better.

    True   False

18. Combination waste and vent systems have vents on each fixture to promote better air flow.

    True   False

19. A trap is useless without water in it.

    True   False

20. Relief vents must be at least one-half the size of the pipes they are venting.

    True   False

21. A hat vent is a vent that extends upward from a trap or trap arm.

    True   False

22. Any building equipped with plumbing must also be equipped with a main vent.

    True   False

23. Vents can be used for more than plumbing, such as supporting antennas, and flagpoles.

    True   False

24. Vent sizing is generally done based on the number of fixtures units entering the stack.

    True   False

25. A typical grade for vent piping is .25 inch to the foot.

    True   False

26.   A trap seal is defined as the vertical distance between the weir and the top of the dip of a trap.
        True   False

27.   An individual vent serves individual fixtures, but they may connect into another vent that extends into the open air.
        True   False

28.   Not all local codes use the same sizing chart, so you should always check the local code before trusting your sizing.
        True   False

29.   Offstacks are not permitted in the stack vent.
        True   False

30.   Wet vents are pipes that serve as a vent for one fixture and a drain for another.
        True   False

MULTIPLE-CHOICE ANSWERS

1. c	19. d	37. b	55. c	73. d
2. a	20. d	38. c	56. a	74. d
3. a	21. c	39. d	57. b	75. b
4. c	22. a	40. b	58. c	76. d
5. c	23. c	41. a	59. d	77. b
6. b	24. b	42. a	60. b	78. c
7. b	25. a	43. b	61. b	79. c
8. d	26. a	44. b	62. b	80. d
9. b	27. d	45. b	63. b	81. c
10. c	28. c	46. b	64. c	82. c
11. b	29. d	47. c	65. b	83. b
12. c	30. d	48. a	66. b	84. a
13. c	31. d	49. d	67. c	85. d
14. b	32. c	50. d	68. a	86. d
15. b	33. a	51. a	69. c	87. b
16. d	34. c	52. c	70. c	88. a
17. b	35. b	53. d	71. c	89. b
18. c	36. b	54. a	72. c	90. b

FILL IN THE BLANK ANSWERS

1. sanitary
2. combination
3. drum traps
4. relief
5. Circuit

6. horizontally
7. vertically
8. a soil pipe
9. atmospheric
10. Island vents

TRUE-FALSE ANSWERS

1. True	9. True	17. True	24. True
2. False	10. True	18. False	25. True
3. True	11. False	19. True	26. True
4. False	12. True	20. True	27. True
5. False	13. False	21. False	28. True
6. True	14. False	22. True	29. False
7. False	15. True	23. False	30. True
8. False	16. True		

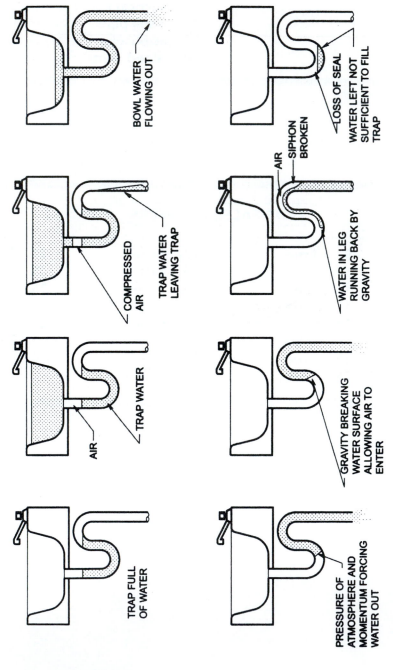

FIGURE 9.1   Self-siphoning phenomenon. *Copyright 2002, International Code Council, Inc., Falls Church, Virginia. Reproduced with permission. All rights reserved.*

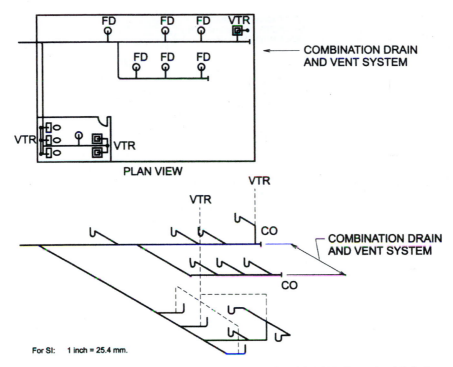

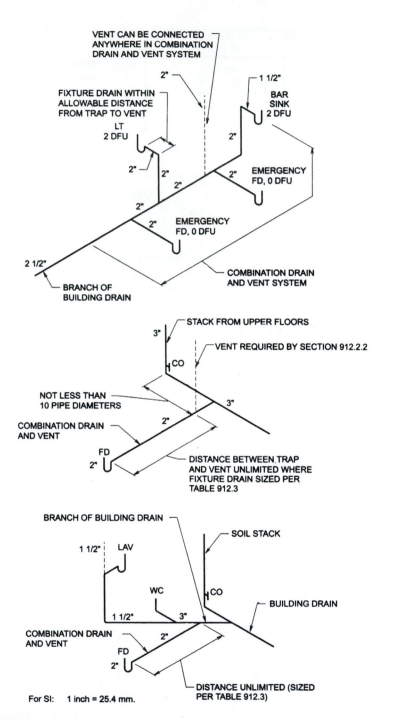

FIGURE 9.3   Combination drain and vent system. *Copyright 2002, International Code Council, Inc., Falls Church, Virginia. Reproduced with permission. All rights reserved.*

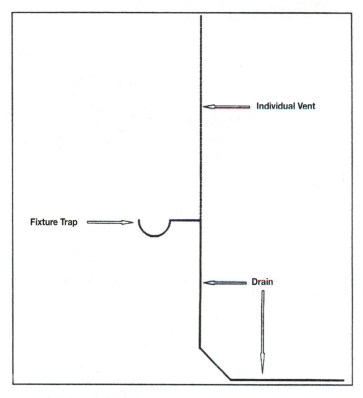

FIGURE 9.4   Individual vent.

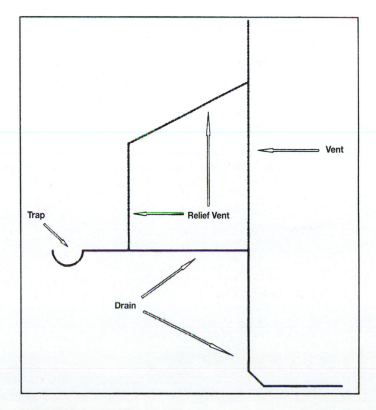

FIGURE 9.5   Relief vent.

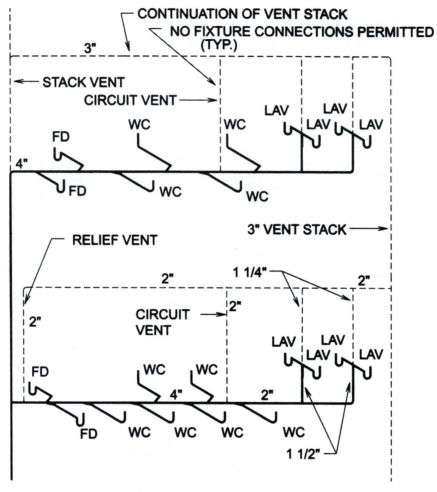

For SI: 1 inch = 25.4 mm.

FIGURE 9.6 Circuit-vented branch with additional fixture connections. *Copyright 2002, International Code Council, Inc., Falls Church, Virginia. Reproduced with permission. All rights reserved.*

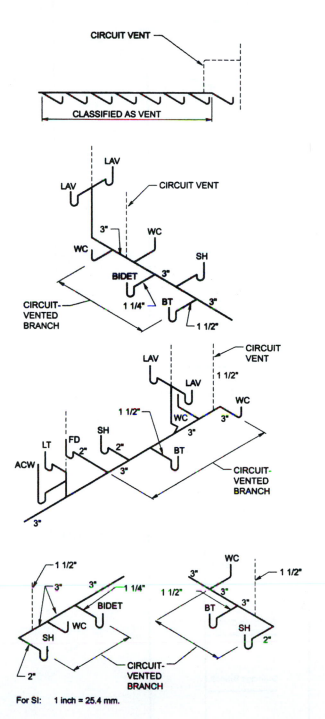

FIGURE 9.7 Circuit venting. *Copyright 2002, International Code Council, Inc., Falls Church, Virginia. Reproduced with permission. All rights reserved.*

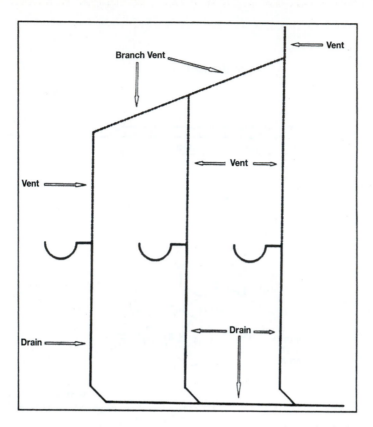

FIGURE 9.8   Branch vent.

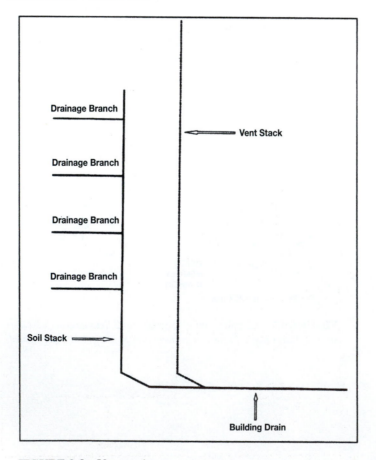

FIGURE 9.9   Vent stack.

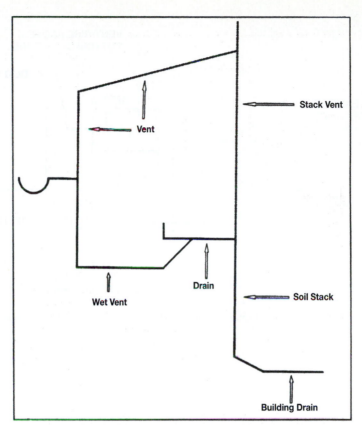

FIGURE 9.10   Stack vent.

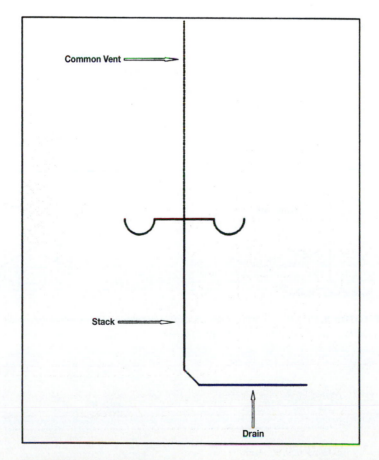

FIGURE 9.11   Common vent.

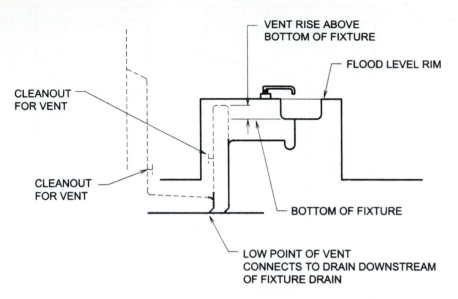

FIGURE 9.12   Island fixture vent. *Copyright 2002, International Code Council, Inc., Falls Church, Virginia. Reproduced with permission. All rights reserved.*

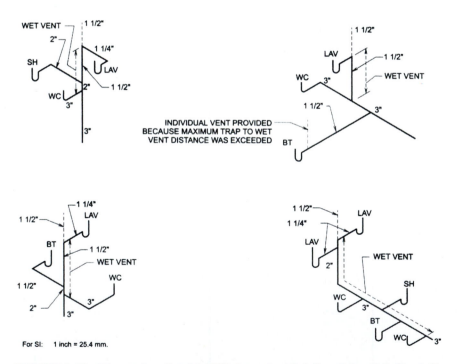

FIGURE 9.13   Wet venting. *Copyright 2002, International Code Council, Inc., Falls Church, Virginia. Reproduced with permission. All rights reserved.*

THE DISTANCE FROM EACH TUB TO POINT B MUST
BE WITHIN THE DISTANCE SPECIFIED IN TABLE 906.1

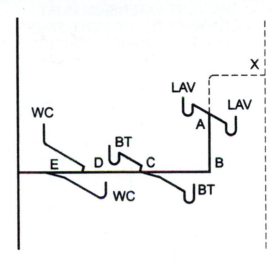

DRAIN COMPONENT	MIN. SIZE	DFU'S (TABLE 709.1)
LAV DRAIN	1 1/4"	1
BT DRAIN	1 1/2"	2
WC DRAIN	3"	4
A-B[a]	2"[a]	2
B-C[a]	2"[a]	2
C-D[a]	2 1/2"[a]	6
D-E[a]	3"[a]	9
A-X	1 1/2"	NA

a = FROM TABLE 909.3
NA = NOT APPLICABLE

For SI: 1 inch = 25.4 mm.

FIGURE 9.14   Double bathroom group (private) wet vent. *Copyright 2002, International Code Council, Inc., Falls Church, Virginia. Reproduced with permission. All rights reserved.*

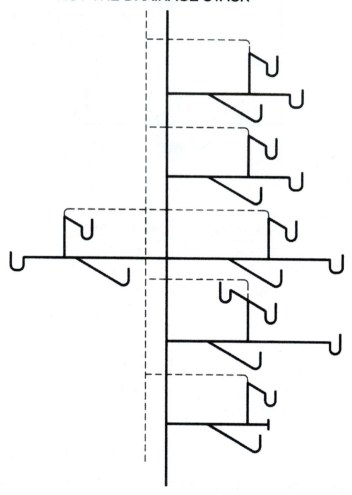

THE VENT EXTENSION MUST
CONNECT TO THE VENT STACK
NOT THE DRAINAGE STACK

FIGURE 9.15 Multifloor wet venting. *Copyright 2002, International Code Council, Inc., Falls Church, Virginia. Reproduced with permission. All rights reserved.*

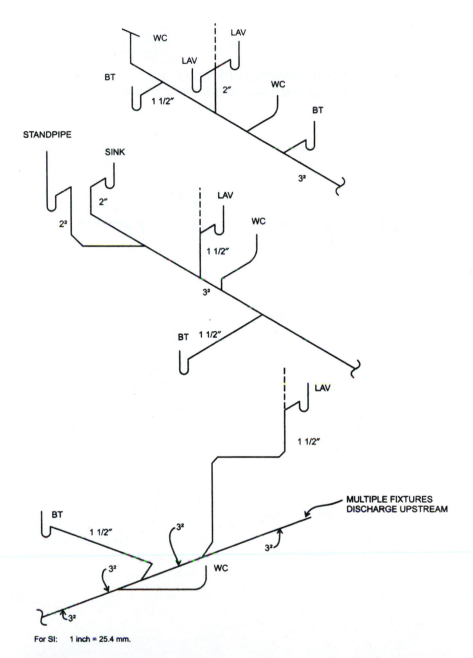

WC
LAV
LAV
BT
1 1/2″
2″
WC
BT
3²
STANDPIPE
SINK
2″
2²
LAV
WC
1 1/2″
3²
BT   1 1/2″
LAV
1 1/2″
BT
1 1/2″
3²
MULTIPLE FIXTURES
DISCHARGE UPSTREAM
3²
WC
3²
3²

For SI: 1 inch = 25.4 mm.

FIGURE 9.16   Prohibited wet venting. *Copyright 2002, International Code Council, Inc., Falls Church, Virginia. Reproduced with permission. All rights reserved.*

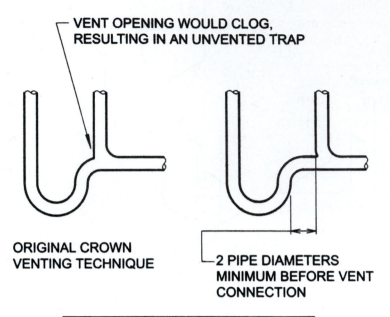

VENT OPENING WOULD CLOG,
RESULTING IN AN UNVENTED TRAP

ORIGINAL CROWN
VENTING TECHNIQUE

2 PIPE DIAMETERS
MINIMUM BEFORE VENT
CONNECTION

TRAP SIZE	MINIMUM DISTANCE TRAP TO VENT
1 1/4"	2 1/2"
1 1/2"	3"
2"	4"
3"	6"

For SI: 1 inch = 25.4 mm.

FIGURE 9.17   Crown venting. *Copyright 2002, International Code Council, Inc., Falls Church, Virginia. Reproduced with permission. All rights reserved.*

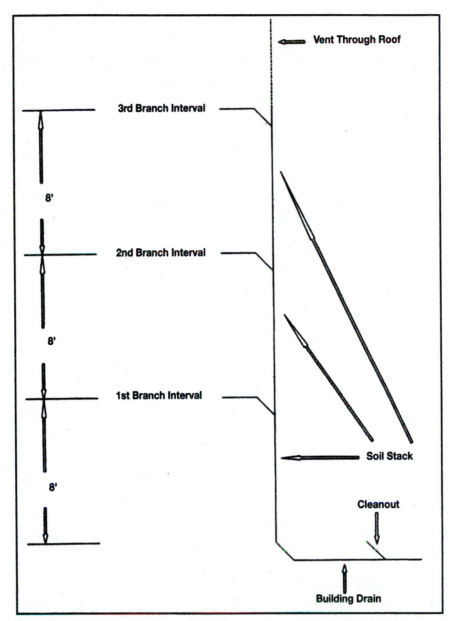

FIGURE 9.18 Branch interval detail.

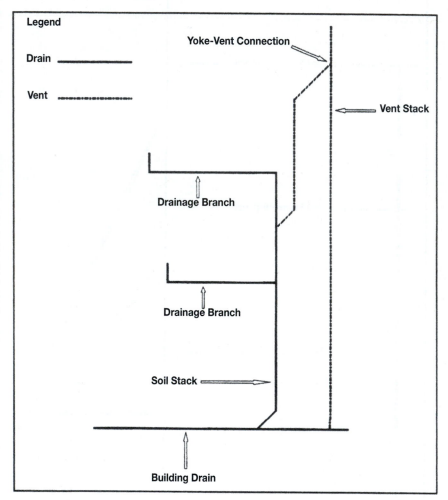

FIGURE 9.19 Yoke vent.

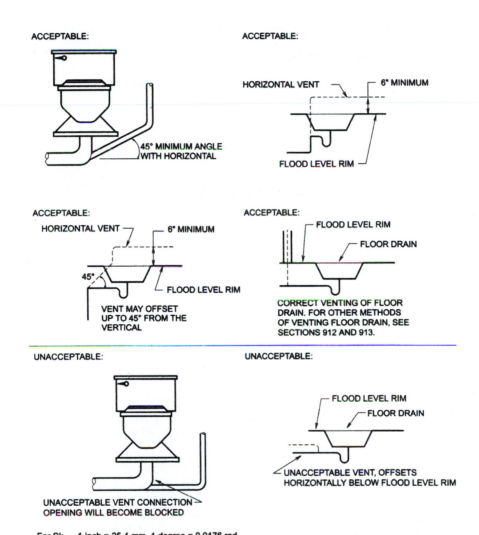

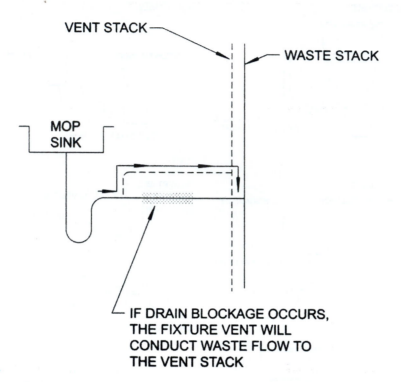

FIGURE 9.21 Improperly connected vent serving a drain. *Copyright 2002, International Code Council, Inc., Falls Church, Virginia, Reproduced with permission. All rights reserved.*

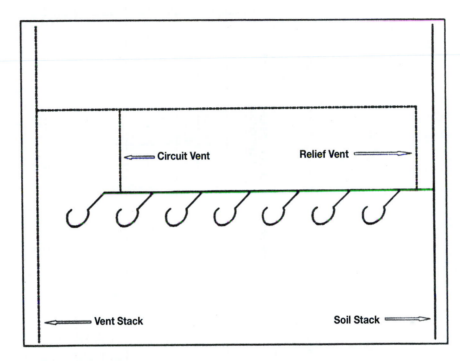

FIGURE 9.22 Circuit vent with relief vent.

TABLE 9.1 Size and developed length of stack vents and vent stacks. *Copyright 2006, International Code Council, Inc., Falls Church, Virginia. Reproduced with permission. All rights reserved.*

DIAMETER OF SOIL OR WASTE STACK (inches)	TOTAL FIXTURE UNITS BEING VENTED (dfu)	MAXIMUM DEVELOPED LENGTH OF VENT (feet)[a] DIAMETER OF VENT (inches)										
		1¼	1½	2	2½	3	4	5	6	8	10	12
1¼	2	30	—	—	—	—	—	—	—	—	—	—
1½	8	50	150	—	—	—	—	—	—	—	—	—
1½	10	30	100	—	—	—	—	—	—	—	—	—
2	12	30	75	200	—	—	—	—	—	—	—	—
2	20	26	50	150	—	—	—	—	—	—	—	—
2½	42	—	30	100	300	—	—	—	—	—	—	—
3	10	—	42	150	360	1,040	—	—	—	—	—	—
3	21	—	32	110	270	810	—	—	—	—	—	—
3	53	—	27	94	230	680	—	—	—	—	—	—
3	102	—	25	86	210	620	—	—	—	—	—	—
4	43	—	—	35	85	250	980	—	—	—	—	—
4	140	—	—	27	65	200	750	—	—	—	—	—
4	320	—	—	23	55	170	640	—	—	—	—	—
4	540	—	—	21	50	150	580	—	—	—	—	—
5	190	—	—	—	28	82	320	990	—	—	—	—
5	490	—	—	—	21	63	250	760	—	—	—	—
5	940	—	—	—	18	53	210	670	—	—	—	—
5	1,400	—	—	—	16	49	190	590	—	—	—	—

(continued)

DIAMETER OF SOIL OR WASTE STACK (inches)	TOTAL FIXTURE UNITS BEING VENTED (dfu)	MAXIMUM DEVELOPED LENGTH OF VENT (feet)[a] DIAMETER OF VENT (inches)										
		1¼	1½	2	2½	3	4	5	6	8	10	12
6	500	—	—	—	—	33	130	400	1,000	—	—	—
6	1,100	—	—	—	—	26	100	310	780	—	—	—
6	2,000	—	—	—	—	22	84	260	660	—	—	—
6	2,900	—	—	—	—	20	77	240	600	—	—	—
8	1,800	—	—	—	—	—	31	95	240	940	—	—
8	3,400	—	—	—	—	—	24	73	190	720	—	—
8	5,600	—	—	—	—	—	20	62	160	610	—	—
8	7,600	—	—	—	—	—	18	56	140	560	—	—
10	4,000	—	—	—	—	—	—	31	78	310	960	—
10	7,200	—	—	—	—	—	—	24	60	240	740	—
10	11,000	—	—	—	—	—	—	20	51	200	630	—
10	15,000	—	—	—	—	—	—	18	46	180	570	—
12	7,300	—	—	—	—	—	—	—	31	120	380	940
12	13,000	—	—	—	—	—	—	—	24	94	300	720
12	20,000	—	—	—	—	—	—	—	20	79	250	610
12	26,000	—	—	—	—	—	—	—	18	72	230	500
15	15,000	—	—	—	—	—	—	—	—	40	130	310
15	25,000	—	—	—	—	—	—	—	—	31	96	240
15	38,000	—	—	—	—	—	—	—	—	26	81	200
15	50,000	—	—	—	—	—	—	—	—	24	74	180

For SI: 1 inch = 25.4 mm, 1 foot = 304.8 mm.

a. The developed length shall be measured from the vent connection to the open air.

TABLE 9.2 Common vent sizes. *Copyright 2006, International Code Council, Inc., Falls Church, Virginia. Reproduced with permission. All rights reserved.*

PIPE SIZE (inches)	MAXIMUM DISCHARGE FROM UPPER FIXTURE DRAIN (dfu)
$1\frac{1}{2}$	1
2	4
$2\frac{1}{2}$ to 3	6

For SI: 1 inch = 25.4 mm.

TABLE 9.3 Wet vent size. *Copyright 2006, International Code Council, Inc., Falls Church, Virginia. Reproduced with permission. All rights reserved.*

WET VENT PIPE SIZE (inches)	DRAINAGE FIXTURE UNIT LOAD (dfu)
$1\frac{1}{2}$	1
2	4
$2\frac{1}{2}$	6
3	12

For SI: 1 inch = 25.4 mm.

TABLE 9.4 Size and length of sump vents. *Copyright 2006, International Code Council, Inc., Falls Church, Virginia.*

DISCHARGE CAPACITY OF PUMP (gpm)	MAXIMUM DEVELOPED LENGTH OF VENT (feet)[a]					
	Diameter of vent (inches)					
	1¼	1½	2	2½	3	4
10	No limit[b]	No limit	No limit	No limit	No limit	No limit
20	270	No limit	No limit	No limit	No limit	No limit
40	72	160	No limit	No limit	No limit	No limit
60	31	75	270	No limit	No limit	No limit
80	16	41	150	380	No limit	No limit
100	10[c]	25	97	250	No limit	No limit
150	Not permitted	10[c]	44	110	370	No limit
200	Not permitted	Not permitted	20	60	210	No limit
250	Not permitted	Not permitted	10	36	132	No limit
300	Not permitted	Not permitted	10[c]	22	88	380
400	Not permitted	Not permitted	Not permitted	10[c]	44	210
500	Not permitted	Not permitted	Not permitted	Not permitted	24	130

For SI: 1 inch = 25.4 mm, 1 foot = 304.8 mm, 1 gallon per minute = 3.785 L/m.

a. Developed length plus an appropriate allowance for entrance losses and friction due to fittings, changes in direction and diameter. Suggested allowances shall be obtained from NSB Monograph 31 or other approved sources. An allowance of 50 percent of the developed length shall be assumed if a more precise value is not available.

b. Actual values greater than 500 feet.

c. Less than 10 feet.

TABLE 9.5 Minimum diameter and maximum length of individual branch fixture vents and individual fixture header vents for smooth pipes. *Copyright 2006, International Code Council, Inc., Falls Church, Virginia. Reproduced with permission. All rights reserved.*

DIAMETER OF VENT PIPE (inches)	INDIVIDUAL VENT AIRFLOW RATE (cubic feet per minute)																			
	Maximum developed length of vent (feet)																			
	1	2	3	4	5	6	7	8	9	10	11	12	13	14	15	16	17	18	19	20
$1/2$	95	25	13	8	5	4	3	2	1	1	1	1	1	1	1	1	1	1	1	1
$3/4$	100	88	47	30	20	15	10	9	7	6	5	4	3	3	3	2	2	2	2	1
1	—	—	100	94	65	48	37	29	24	20	17	14	12	11	9	8	7	7	6	6
$1^{1}/_{4}$	—	—	—	—	—	—	—	100	87	73	62	53	46	40	36	32	29	26	23	21
$1^{1}/_{2}$	—	—	—	—	—	—	—	—	—	—	—	100	96	84	75	65	60	54	49	45
2	—	—	—	—	—	—	—	—	—	—	—	—	—	—	—	—	—	—	—	100

For SI: 1 inch = 25.4 mm, 1 cubic foot per minute = 0.4719 L/s, 1 foot = 304.8 mm.

TABLE 9.6   Maximum distance of fixture trap from vent.
*Copyright 2006, International Code Council, Inc., Falls Church, Virginia.
Reproduced with permission. All rights reserved.*

SIZE OF TRAP (inches)	SLOPE (inch per foot)	DISTANCE FROM TRAP (feet)
$1\frac{1}{4}$	$\frac{1}{4}$	5
$1\frac{1}{2}$	$\frac{1}{4}$	6
2	$\frac{1}{4}$	8
3	$\frac{1}{8}$	12
4	$\frac{1}{8}$	16

For SI: 1 inch = 25.4 mm, 1 foot = 304.8 mm,
1 inch per foot = 83.3 mm/m.

TABLE 9.7   Waste stack vent size. *Copyright 2006, International Code
Council, Inc., Falls Church, Virginia. Reproduced with permission. All rights
reserved.*

STACK SIZE (inches)	MAXIMUM NUMBER OF DRAINAGE FIXTURE UNITS (dfu)	
	Total discharge into one branch interval	Total discharge for stack
$1\frac{1}{2}$	1	2
2	2	4
$2\frac{1}{2}$	No limit	8
3	No limit	24
4	No limit	50
5	No limit	75
6	No limit	100

For SI: 1 inch = 25.4 mm.

Chapter 10

TRAPS, CLEAN-OUTS, AND INTERCEPTORS

This chapter is going to test your knowledge of traps, interceptors, and separators. While traps are not complicated devices, complying with the plumbing code gives some plumbers trouble. There is sometimes doubt as to whether an S-trap can be replaced with another S-trap or whether a different type of trap is required. Drum traps are common in Maine and rare in Virginia. Why is this? The answer is that different venting systems are required in the two states, but there are many circumstances that can affect the trap that is used with a particular fixture.

A lot of plumbers become confused when asked what the difference is between a trap and an interceptor. In fact, some plumbers don't even know what separators are. It's true that separators are not used on a regular basis by average plumbers, but that doesn't mean that a plumber is not required to understand them for the licensing exam.

Do you know when a separator should be installed? Do you know what it is? Does a separator serve a water-distribution system or a drainage system? If you can't answer all these questions correctly at this time, don't worry; this chapter will enable you to learn about and understand these terms.

Almost any plumber with a little experience knows what a P-trap is, but a much smaller percentage of plumbers can describe what a crown-vented trap is. Do you know? How about a bell trap—can you describe what it is or where it is used? Are S-traps legal in new installations?

As you can see, there are many questions that can come up in regard to traps, and we've barely scratched the surface. Since all plumbing fixtures are usually trapped, it makes sense that knowing which traps to use, when to use them, and where to install them is an important part of plumbing.

The subject of traps, interceptors, and separators doesn't take up a lot of space in a code book, and it won't consume a large number of pages in this book. This doesn't mean, however, that you should look past this chapter and move on to more interesting topics.

Traps, interceptors, and separators play important roles in plumbing systems, and they deserve your attention and respect. There could be enough questions on your licensing exam that deal with these elements of a plumbing system to have an adverse affect on your overall score if you are not knowledgeable on the topic.

I know from past experience, both as a field plumber and a code-class instructor, that apprentices, journeymen, and even master plumbers tend to pay little attention to the section of the plumbing code that deals with traps, interceptors, and separators. Some of my past students have felt that this section of the code was too simple to be concerned about. After taking some of my mock exams, they changed their minds. If you have any preconceived ideas about how simple the code is in regard to these issues, clear your mind and prepare to learn.

Many potential plumbers fail to understand that an exam can be worded to make a simple subject appear complicated. Some plumbing exams can have the best plumbers scratching their heads. I'm warning you: don't take any part of the plumbing code for granted. If you make a mistake in the field, your master plumber or the local plumbing inspector will probably catch it and have you correct it. If you make too many mistakes on your licensing exam, you will not get the license that you need to enjoy better working conditions and more money. Treat every question on your exam as if it stood between you and your plumber's license. Now, let's move on to the multiple-choice exam for this chapter.

MULTIPLE-CHOICE EXAM

1. A device, fitting, or assembly of fittings installed in a building drain for the purpose of preventing air circulation between the drainage system of the building and the building sewer is known as:

    a. A backwater valve   b. An interceptor

    c. A separator   d. A building trap

2. A device, fitting, or assembly of fittings installed to provide a liquid seal to prevent the emission of sewer gas without materially affecting the flow of sewage or wastewater through the device is known as:

    a. An air chamber   b. A trap

    c. A gas interceptor   d. An approved obstruction

3. The vertical distance between the crown weir and the top of the dip of a trap is known as:

    a. An airgap   b. An airbreak

    c. The critical level   d. The trap seal

4. A device that is designed and installed to separate and retain for removal deleterious, hazardous, or undesirable matter from normal wastes while permitting normal sewage or wastes to discharge into the drainage system by gravity is known as:

    a. A backwater valve   b. An interceptor

    c. A drum trap   d. A screener

5. Excluding exceptions, every plumbing fixture must be equipped with:

    a. Cutoff valves   b. A water-sealed trap

    c. A gas-sealed trap   d. A P-trap

6. When a trap is installed to serve a plumbing fixture, the trap must be installed as close as possible to:

    a. The floor   b. The trap arm

    c. The vent   d. The fixture outlet

7. The vertical distance from the fixture outlet to the inlet of a grease trap that is being used as a fixture trap must not exceed:

    a. 6 inches   b. 12 inches

    c. 24 inches   d. 30 inches

8. No plumbing fixture is allowed to be:

    a. Served by a continuous waste

    b. Trapped with a drum trap

    c. More than 20 inches from its vent

    d. Double-trapped

9. The trap of a kitchen sink may not accept the waste of:

    a. A dishwasher   b. A garbage disposer

    c. A continuous waste   d. A clothes washer

10. The drainage discharge from a laundry tub must not discharge into:

    a. A P-trap   b. A sanitary drainage system

    c. A septic tank   d. A kitchen-sink trap

11. Combination plumbing fixtures, such as double-bowl kitchen sinks, are allowed to share a single trap when none of the compartments of the sink bowls are more than _____ inches deeper than the others. Fill in the blank:

    a. 4   b. 6

    c. 8   d. 12

12. Combination plumbing fixtures, such as double-bowl kitchen sinks, are allowed to share a single trap when the waste outlets are not more than _____ inches apart. Fill in the blank:

    a. 12   b. 15

    c. 30   d. 36

13. Fixture traps must not have:

    a. Plastic bodies   b. Slip-joint connections

    c. Interior partitions   d. Rough-brass finishes

14. Drum traps are prohibited, except:

    a. For use with lavatories   b. For use with floor drains

    c. For use with condensate drains   d. Where approved

15. Traps must be set level with respect to the:

    a. Position of the fixture being served

    b. Position of the floor beneath the trap

    c. Trap seal

    d. Direction of flow

16. Which of the following types of traps are prohibited:

a. S-traps
b. Crown-vented traps
c. All of the above
d. None of the above

17. Which of the following types of traps are prohibited:

a. P-traps
b. Crown-vented traps
c. All of the above
d. None of the above

18. Which of the following types of traps are prohibited:

a. Bell traps
b. Traps that depend upon moving parts to maintain a seal
c. All of the above
d. None of the above

19. Which of the following types of traps are prohibited:

a. Separate fixture traps that depend on interior partitions for the seal
b. Traps that depend upon moving parts to maintain a seal
c. All of the above
d. None of the above

20. Each fixture trap is required to have a liquid seal with a minimum depth of:

a. 2 inches
b. 3 inches
c. 4 inches
d. 6 inches

21. Each fixture trap is required to have a liquid seal with a maximum depth of:

a. 2 inches
b. 3 inches
c. 4 inches
d. 6 inches

22. Traps installed in unheated areas must be protected against:

a. Condensation
b. Freezing
c. Vandalism
d. All of the above

23. Under normal conditions, which of the following cannot be installed without permission from a code official:

a. Drum traps
b. Building traps
c. Both A and B
d. None of the above

24. When a building trap is installed, it must be provided with:
    a. A cleanout
    b. A relief vent or fresh-air intake
    c. Both A and B
    d. Either A or B

25. When an acid-resisting trap is installed underground, the trap must be embedded in concrete that extends _____ inches beyond the bottom and sides of the trap. Fill in the blank:
    a. 2
    b. 4
    c. 6
    d. 12

26. When a building trap is installed and a relief vent is provided for it, the vent must have a diameter that is at least _____. Fill in the blank:
    a. Not less than one-half the size of the drain it is connected to
    b. Equal to the size of the drain it is connected to
    c. 3 inches in size
    d. None of the above

27. Which of the following substances warrant the use of an interceptor:
    a. Oil
    b. Grease
    c. Sand
    d. All of the above

28. Grease interceptors are not required for:
    a. Motel kitchens
    b. Individual dwelling units
    c. Correctional facilities
    d. Restaurants

29. Grease interceptors are not required for:
    a. Private living quarters
    b. Individual dwelling units
    c. Both A and B
    d. Correctional facilities

30. A fresh-air intake or vent for a building trap must be:
    a. Carried above grade
    b. 3 inches in diameter
    c. Made of plastic pipe
    d. None of the above

31. Which of the following locations require the installation of a separator:
    a. An apartment building
    b. A bank building
    c. A single-family residence
    d. A repair garage with grease racks

32. Grease interceptors must be equipped with devices to control the rate of:

a. Air flow

b. Water flow

c. Heat transference

d. All of the above

33. Oil separators must have a depth of at least _____ feet below the invert of the discharge drain. Fill in the blank:

a. 2

b. 4

c. 5

d. 6

34. The outlet opening of an oil separator must have a water seal of not less than _____ inches. Fill in the blank:

a. 18 inches

b. 22 inches

c. 24 inches

d. 36 inches

35. Separators that are installed in areas where automobiles are serviced must have a minimum capacity of _____ cubic feet for the first 100 square feet of area to be drained. Fill in the blank:

a. 4

b. 6

c. 8

d. 12

36. Separators, as described in question number 35, must be enlarged at a rate of _____ cubic inches for each additional 100 square feet of area to be drained. Fill in the blank:

a. 4

b. 6

c. 8

d. 12

37. Sand interceptors must be installed so that they are:

a. Accessible

b. Readily accessible

c. Concealed

d. None of the above

38. Sand interceptors must have water seals with depths of not less than:

a. 6 inches

b. 10 inches

c. 14 inches

d. 18 inches

39. Filters used in commercial laundries must be sized to prohibit the entrance of articles with diameters of _____ or larger from entering the drainage system. Fill in the blank:

a. 1/4 inch

b. 1/2 inch

c. 3/4 inch

d. 1 inch

40.   Commercial laundries must be equipped with an interceptor with a _____ or similar device, which is removable for cleaning. Fill in the blank:

    a.   Water seal   b.   Gas seal

    c.   Wire basket   d.   Brush

41.   The interceptors used in bottling establishments must be capable of separating _____ from other solids before discharging the waste into the drainage system. Fill in the blank:

    a.   Flammable materials   b.   Broken glass

    c.   Oil   d.   Bottled fluids

42.   Separators used in slaughterhouses must be capable of preventing _____ from entering the drainage system. Fill in the blank:

    a.   Feathers   b.   Entrails

    c.   Both A and B   d.   None of the above

43.   If an interceptor or separator is subject to the loss of a trap seal, the device must be:

    a.   Drained of all air   b.   Vented

    c.   Inspected regularly   d.   All of the above

44.   Interceptors must be cleaned:

    a.   Periodically   b.   Every Monday

    c.   At the end of each work week   d.   Once a month

45.   Separators must be cleaned:

    a.   Periodically   b.   Every week

    c.   At the end of each work week   d.   Once a month

You have now completed the multiple-choice testing on traps, interceptors, and separators. Were the questions a bit more complex that you expected? I suspect they were. As you can see, there is quite a bit to be learned about these devices. If you underestimated the importance of traps, interceptors, and separators, be glad you did so on this mock examination rather than on your licensing exam. Now, proceed directly to the fill in the blank and the true-false test for your region and continue your testing.

FILL IN THE BLANK EXAM

---

1.   Generally, clean-outs are required where the _____ drain meets the building sewer.

2.   Clean-outs are usually required every time a sewer turns more than _____ degrees.

3.   _____ traps are not allowed for use in new installations.

4.   The tailpiece between a fixture drain and the fixture's trap may not exceed _____ inches.

5.   _____ are designed to control what goes down a drain.

6.   All _____ must have clean-outs.

7.   No under-floor clean-out is allowed to be placed more than _____ feet from an access opening.

8.   Plugs used for clean-outs are to be constructed of plastic or _____.

9.   Backwater valves are installed in _____ and sewers to prevent water and waste backup.

10.   Some jurisdictions prefer that the clean-outs at the junction of building drains and sewers be located _____.

Cross off the answers as you use them

| Interceptors | 24 | sewers | bell | 20 | 45 |

| brass building | drains | brass | outside | | |

TRUE-FALSE EXAM

1.   All plumbing fixtures, except those with integral traps, must be trapped with an approved, water-seal trap.

     True   False

2.   All commercial garbage disposers must be trapped individually.

     True   False

3.   An automatic clothes washer can only discharge into the trap of a kitchen sink when the clothes washer is installed adjacent to the sink.

     True   False

4.   The vertical distance between a fixture outlet and the trap weir cannot exceed 30 inches.

     True   False

5.   Every plumbing trap, excluding exceptions [such as?], must be vented.

     True   False

6.   All traps, except those used for interceptors and similar devices, must be self-cleaning.

     True   False

7.   The trap for a fixture drain shall not be larger than the trap arm it is connected to.

     True   False

8.   S-traps are prohibited.

     True   False

9.   Bell traps are allowed only for use with floor drains.

     True   False

10. Crown-vented traps are prohibited.

     True   False

11. An approved two-way clean-out is not allowed in locations where a building drain meets a sewer drain.

     True   False

12. When a P-trap is allowed as a clean-out, it may be smaller than the drain.

     True   False

13. No fixture is allowed to be double-trapped.

True   False

14. Every trap for every fixture is required to have a trap seal that is made with a liquid.

True   False

15. P-traps are the least used traps in modern plumbing systems.

True   False

16. Any type of trap may be used at any time without special permission from the code officer.

True   False

17. Hat-vented traps have a vent rising from the top of the trap.

True   False

18. Traps must be installed level in order for the trap seal to function properly.

True   False

19. There is a big difference between a trap and an interceptor.

True   False

20. Oil separators are not required at repair garages and gas stations, but are recommended by code officials.

True   False

! Code alert

Fixtures and equipment to include:

1. Pot sinks
2. Pre-rinse sinks
3. Soup kettles or similar devices
4. Wok stations
5. Floor drains
6. Sinks into which kettles are drained
7. Automatic hood wash units
8. Dishwashers without prerinse sinks

Grease interceptors and automatic grease removal devices will only receive wastes from fixtures and equipment that allow fats, oils, or grease discharge. Emulsifiers, chemicals, enzymes, and bacteria must not discharge into the food waste grinder.

21.   Clean-outs are required to be the same size as the pipe they are serving unless the pipe is larger than six inches.

True   False

22.   If a drain is less than 5 feet long and is not used for sinks or urinals, a clean-out is not required.

True   False

23.   Clean-outs are frequently required at the base of every stack.

True   False

24.   The ultimate clean-out is a manhole.

True   False

25.   A trap arm is the section of pipe that extends from a fixture drain's trap to a drain.

True   False

MULTIPLE-CHOICE ANSWERS

1. d	10. d	19. c	28. b	37. b
2. b	11. b	20. a	29. c	38. a
3. d	12. c	21. c	30. a	39. b
4. b	13. c	22. c	31. d	40. c
5. b	14. d	23. a	32. b	41. b
6. d	15. c	24. c	33. b	42. c
7. d	16. c	25. c	34. a	43. b
8. d	17. b	26. a	35. b	44. a
9. d	18. c	27. d	36. d	45. a

FILL IN THE BLANK ANSWERS

1.	Building	6.	Sewers
2.	45	7.	20
3.	Bell	8.	Brass
4.	24	9.	Drains
5.	Interceptors	10.	Outside

TRUE-FALSE ANSWERS

1. True	7. True	13. True	19. True
2. True	8. True	14. True	20. False
3. False	9. False	15. False	21. False
4. False	10. True	16. False	22. False
5. True	11. False	17. False	23. True
6. True	12. True	18. True	24. True
			25. True

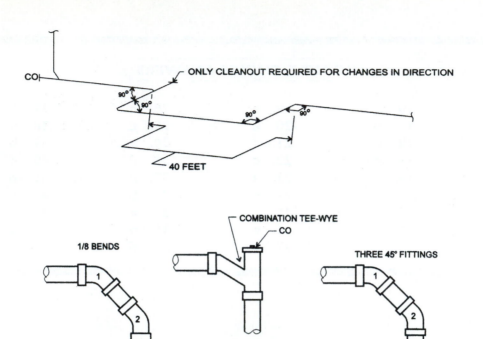

FIGURE 10.1   Examples of clean-out locations. *Copyright 2002, International Code Council, Inc., Falls Church, Virginia. Reproduced with permission. All rights reserved.*

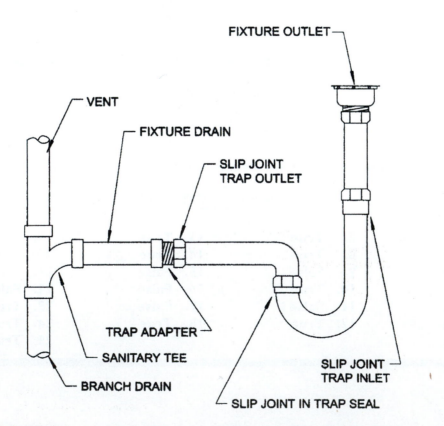

FIGURE 10.2   A P-trap, like this one, can be used as a clean-out. *Copyright 2002, International Code Council, Inc., Falls Church, Virginia. Reproduced with permission. All rights reserved.*

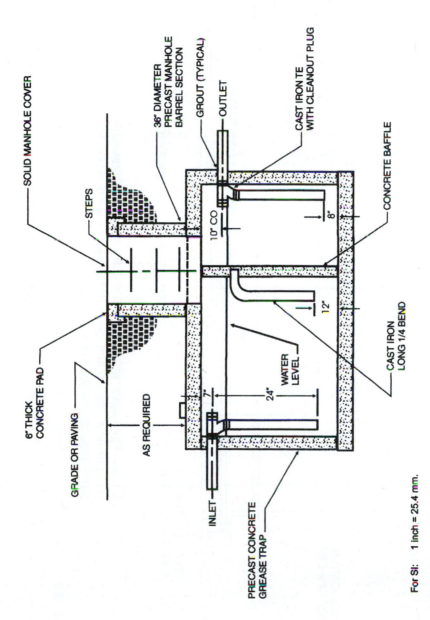

FIGURE 10.3 Typical concrete grease interceptor. *Copyright 2002, International Code Council, Inc., Falls Church, Virginia. Reproduced with permission. All rights reserved.*

For SI: 1 inch = 25.4 mm.

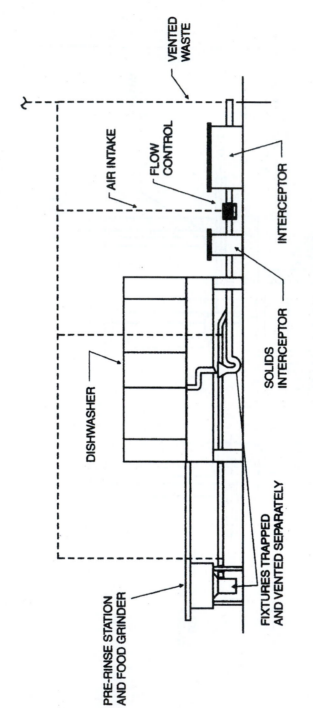

FIGURE 10.4 Venting a grease trap. *Copyright 2002, International Code Council, Inc., Falls Church, Virginia. Reproduced with permission. All rights reserved.*

TABLE 10.1   Capacity of grease interceptors*a*. *Copyright 2006, International Code Council, Inc., Falls Church, Virginia. Reproduced with permission. All rights reserved.*

TOTAL FLOW-THROUGH RATING (gpm)	GREASE RETENTION CAPACITY (pounds)
4	8
6	12
7	14
9	18
10	20
12	24
14	28
15	30
18	36
20	40
25	50
35	70
50	100
75	150
100	200

For SI: 1 gallon per minute = 3.785 L/m, 1 pound = 0.454 kg.

a. For total flow-through ratings greater than 100 (gpm) , double the flow-through rating to determine the grease retention capacity (pounds).

Chapter 11
STORM DRAINAGE

The basics of storm-water drainage systems are easy to understand. These types of drainage systems are simply utilized to control the buildup and runoff of rainwater. This, of course, is oversimplifying the purpose of storm-water drainage, but it is the most basic definition of the systems.

Computing and designing storm water systems can become quite complicated. The math alone can be intimidating. However, as a potential master plumber, you will have to learn how to design storm-water systems. This job is sometimes left to engineers and architects, but your licensing exam will test your knowledge of the subject.

The installation of a drainage system to convey storm water is not the same as that of a system meant to transfer sewage and wastewater. Many of the principles are the same, but at the same time there are distinct differences. For example, roof drains are installed for storm water and never wastewater. Some locations allow the commingling of storm water and wastewater, but other jurisdictions prohibit the combined use of a sewer system. There are, to be sure, many differences between sanitary drainage systems and storm-water drainage systems.

In talking about a sanitary drainage system, a vertical pipe that conveys waste from one story to another is called a stack. If you are installing a drainage system for storm water, a similar pipe is called a leader or a conductor, depending upon where it is located.

Subsoil drains are yet another difference when dealing with storm water. These are drains that collect subsurface water and convey it to a point of disposal. There are no such drains in a sanitary drainage system. Area drains are another example of how storm water differs from sanitary drainage. An area drain is similar, in terms of plumbing, to a floor drain, but it is not the same thing.

If you were given a detailed set of blueprints for a commercial building and asked to label all the drains as either sanitary or storm drains, the task could seem very difficult if you don't understand the differences between the two types of drainage. What appears to be a floor drain in an open outside stairway will actually be an area drain. This type of confusion could cost you valuable points on the scoring of your licensing exam.

Licensing exams for plumbers vary in complexity and content. In fact, they are often changed from time to time to prevent the possibility of cheating. Since I cannot predict with absolute certainty what will be on your licensing exam, I will test you on a large part of the material that you might encounter. In all probability, the tests that I give you in this chapter will be more difficult than what you will encounter on your licensing exam. Of course, there is no way of knowing exactly what you will face on the day of your test, so it is better to be over-prepared than to be underprepared.

I'd like to take a moment to tell you about my personal preparation for my first licensing exam. When I got my first job as a plumber's helper, I didn't know anything about plumbing. My first job lasted less than two weeks. I was fired for lack of knowledge.

When I got my second job as a helper, I knew a little more than I did on my first job, but not much more. I still didn't know the difference between a one-eighth bend and a one-sixteenth-bend. The second job only lasted a few weeks before I was terminated. All the plumbing companies at that time seemed to want personnel who were ready to work and be productive. I wanted to do a good job, but I didn't have the knowledge to hold my own. Then I got my third job as a helper.

My third attempt at becoming a plumber was much more successful than my first two attempts had been. By the time I got the third job, I had learned what most of the fittings were and how to tell the difference among various pipe sizes without measuring them. This was a great accomplishment for a green kid. I still didn't know how to solder, but I was improving.

My third plumbing employer was willing to train me. Boy, was this a relief. I learned all the fittings, and I learned how to solder. Most of my first six months on the job were spent digging ditches and running a jackhammer, but every now and then I would get to help with some real plumbing. I thought at least three times about quitting the job due to the excessive physical labor. I hung in there. During those first six months I lost about 60 pounds and gained a new self-discipline.

After my six months of probation were over, things got better. The company allowed me to start cleaning copper fittings and drilling holes. To many this would seem like menial work, but to me it was a sure sign that I would become a plumber. When I cleaned the copper fittings, I would pay attention to how the plumbers used them. As I drilled the holes, I asked questions and learned why the holes were being drilled in specific places. It was about this time that I spent my own money to buy a code book.

I generally arrived at work early each day, and I would read my code book while I waited for the shop to open. My code book was with me at all times while I was on the job. I would read it during lunch and on the few occasions when the plumber decided to drive. As you might imagine, there were a great number of aspects of the code that I couldn't understand. When the timing was right, I would question the plumber I worked with on various points of the code. Fortunately, Jerry spent the time to explain it to me. I guess I owe a great deal to Jerry. You see, without his help, I might not have become the plumber I am today.

In less than a year, I progressed from a know-nothing kid to an advanced plumber's helper. I could run water pipe, solder joints, set fixtures, and do many of the jobs that plumbers normally do. My progression into the trade moved very quickly, and I obtained my master's license at a very early age. Danny, my supervisor, and Jerry, my plumber, both played a role in grooming me to be a great plumber. However, even with their best efforts, I would not have made it without the self-determination to be better than average. While most helpers were goofing off, I was studying the code book. By the time I had the field experience needed to be a plumber, I knew the code inside and out.

Well, you've just indulged me in remembering the past and how I got started in plumbing. Thank you; it is a pleasurable memory for me. I didn't tell you the story just to amuse myself, however; I wanted to make a point. Even as a helper, I thrived on learning and understanding the plumbing code. It is my firm belief that my dedication to the code allowed me to move up the ladder faster than most helpers and journeymen plumbers.

Right now you may not have any interest in learning more than you have to in order to pass your licensing exam. That is your choice, but I'll tell you this: you can never know too much about plumbing. It's been over 20 years since Jerry first showed me how to solder a joint, and I still learn new things on a regular basis in the field of plumbing. The more you assert yourself, the better you will do in life. With that said, let's get on with the multiple-choice exam.

MULTIPLE-CHOICE EXAM

1. A receptacle installed to collect and drain storm water from an open area is called:
    a. A floor drain
    b. An area drain
    c. A split drain
    d. A combination sewer

2. A building drain that conveys storm water but no sewage is called:
    a. A sanitary drain
    b. A surface drain
    c. A storm drain
    d. A combined drain

3.   A drain that collects water from beneath the surface and conveys it to a point of disposal is called:

   a.   A combined drain   b.   A gray-water drain

   c.   An underground drain   d.   A subsoil drain

4.   Which of the following is required in the installation of a storm-drainage system:

   a.   45-degree bends   b.   90-degree bends

   c.   Cleanouts   d.   Blind plugs

5.   Which of the following commonly carries both storm water and sewage:

   a.   A combined sewer   b.   A sanitary sewer

   c.   A mixed drain   d.   A subsoil drain

6.   Which of the following is not allowed in the construction of a storm-water drainage system:

   a.   Drainage fittings that are commonly used for sanitary drainage

   b.   Schedule-40 plastic fittings

   c.   Fittings that retard the flow of storm water

   d.   Cast-iron fittings

7.   Exterior pipes used to convey storm water from a roof drain to a point of disposal are known as:

   a.   Stacks   b.   Wet stacks

   c.   Leaders   d.   Conductors

8.   Interior pipes used to convey storm water from a roof drain to a point of disposal are known as:

   a.   Stacks   b.   Wet stacks

   c.   Leaders   d.   Conductors

9.   Roof drains are designed to discharge storm water into:

   a.   Leaders   b.   Conductors

   c.   Either A or B   d.   None of the above

10. A combined sewer is one that conveys:

   a.   Storm water   b.   Sewage

   c.   Waste   d.   All of the above

11. Under normal conditions, horizontal drains for storm water should have a minimum pitch of _____ inch per foot. Fill in the blank.

   a.   1/16   b.   1/8

   c.   1/4   d.   1/2

12. When a building subdrain is installed below the level of a public sewer, the subdrain must discharge into:

   a. A septic tank   b. A sump

   c. A receiving tank   d. Either B or C

13. Conductors shall not be used as:

   a. Soil pipes   b. Waste pipes

   c. Vent pipes   d. Any of the above

14. Sump pits may be constructed of:

   a. Tile   b. Concrete

   c. Steel   d. Any of the above

15. Sump pits may be constructed of:

   a. Plastic   b. Concrete

   c. Steel   d. All of the above

16. Sumps for subsoil drains are not required to:

   a. Have gas-tight covers   b. Be vented

   c. Both A and B   d. None of the above

17. A sump-pump system consists of:

   a. A sump pump   b. A sump pit

   c. Discharge piping   d. All of the above

18. Sump pits must be fitted with:

   a. Gas-tight covers   b. Removable covers

   c. Hinged covers   d. Air-tight covers

19. When a subsoil drain is subject to backflow, it must be protected with:

   a. A vacuum breaker   b. A gate valve

   c. A backwater valve   d. Any of the above

20. Subsoil drains may discharge into:

   a. A trapped area drain   b. A sump

   c. Either A or B   d. None of the above

21. Subsoil drains may discharge into:

   a. A dry well   b. A sump

   c. Either A or B   d. None of the above

22. When a backwater valve is installed, it must be provided with:

   a. Access   b. A gate valve

   c. A check valve   d. Both B and C

23. A controlled-flow-system design is based on:

   a. The size of the sanitary sewer

   b. The size of the storm sewer

   c. The local rainfall rate

   d. All of the above

24. Discharge piping and fittings installed on sump-pump systems must be:

   a. Nylon

   b. Plastic

   c. The same size or larger than the discharge tapping of the sump pump

   d. The same size or smaller than the discharge tapping of the sump pump

25. The floor of a sump pit must provide:

   a. A minimum area of 10 square feet

   b. A minimum area of 8 square feet

   c. A permanent support for a sump pump

   d. Both B and C

FILL IN THE BLANK EXAM

1.   The first step to take when sizing a storm drain or sewer is to establish your known _____.

2.   You should consider the _____ of a horizontal pipe when sizing a storm-water drainage system.

3.   Storm-water piping requires the same amount of _____, with the same frequency, as a sanitary system.

4.   _____ valves installed in a storm drainage system must conform to local code requirements.

5.   Storm-water and sanitary systems should not be _____.

6.   When rain leaders and _____ drains are allowed to connect to a sanitary sewer, they are required to be trapped.

7.   Traps must be _____ for cleaning.

8.   The lid on a sump pump must be _____.

9.   Construction of a sump pit may be accomplished with _____ iron.

10.   All sump-pump _____ pipes should be equipped with a check valve.

Cross off the answers as you use them

Backwater	criteria	storm	cast	pitch	removable
cleanouts	discharge	combined	accessible		

TRUE-FALSE EXAM

1. A suitable location for a storm drainage piping might be a catch basin.
    True   False

2. Your code book should have all of the key elements required to size a storm-water system.
    True   False

3. In sizing storm-water drainage system one consideration is the pitch of a horizontal pipe.
    True   False

4. Sizing gutters and sizing horizontal storm drains are done in many different ways.
    True   False

5. Roof drains should be at least half the size of the piping connected to them.
    True   False

6. All roofs that cannot drain to hanging gutters are required to have roof drains.
    True   False

7. Approved materials can differ from one jurisdiction to another.
    True   False

8. Some roof designs require a backup drainage system in case of emergencies.
    True   False

9. Storm-water drainage is designed to be piped into a sanitary sewer or plumbing system.
    True   False

10. When working with a standard table like those in most codebooks, they must be converted into data to suit your local conditions.
    True   False

11. You should be able to obtain rainfall figures from your state or county offices
    True   False

12. A surface area should include both roof and parking areas.
    True   False

13. To size a vertical storm drain or sewer, you need to decide what pitch you will put on the pipe.
    True   False

14. Roof drains should be sealed to prevent water from leaking around them.

    True   False

15. You must always protect storm-water piping; they are just as important as plumbing pipes.

    True   False

Now that you've completed the exams, how do you feel about your knowledge of storm-water drainage? You have completed all testing in this chapter for storm water. Check your answers to see how you did on the test.

MULTIPLE-CHOICE ANSWERS

1. b	6. c	11. c	16. d	21. c
2. c	7. c	12. d	17. d	22. a
3. d	8. d	13. d	18. b	23. c
4. c	9. c	14. d	19. c	24. c
5. a	10. d	15. d	20. c	25. c

FILL IN THE BLANK ANSWERS

1. criteria	6. storm
2. pitch	7. accessible
3. clean-outs	8. removable
4. Backwater	9. cast
5. combined	10. discharge

TRUE-FALSE ANSWERS

1. True	5. False	9. False	13. False
2. True	6. True	10. True	14. True
3. True	7. True	11. True	15. True
4. False	8. True	12. True	

TABLE 11.1   Sizing chart for semicircular roof gutters. *Copyright 2006, International Code Council, Inc., Falls Church, Virginia. Reproduced with permission. All rights reserved.*

| DIAMETER OF GUTTERS (inches) | HORIZONTALLY PROJECTED ROOF AREA (square feet) | | | | | |
| | Rainfall rate (inches per hour) | | | | | |
	1	2	3	4	5	6
	$1/16$ unit vertical in 12 units horizontal (0.5-percent slope)					
3	680	340	226	170	136	113
4	1,440	720	480	360	288	240
5	2,500	1,250	834	625	500	416
6	3,840	1,920	1,280	960	768	640
7	5,520	2,760	1,840	1,380	1,100	918
8	7,960	3,980	2,655	1,990	1,590	1,325
10	14,400	7,200	4,800	3,600	2,880	2,400
	$1/8$ unit vertical 12 units horizontal (1-percent slope)					
3	960	480	320	240	192	160
4	2,040	1,020	681	510	408	340
5	3,520	1,760	1,172	880	704	587
6	5,440	2,720	1,815	1,360	1,085	905
7	7,800	3,900	2,600	1,950	1,560	1,300
8	11,200	5,600	3,740	2,800	2,240	1,870
10	20,400	10,200	6,800	5,100	4,080	3,400

			1/4 unit vertical in 12 units horizontal (2-percent slope)			
3	1,360	680	454	340	272	226
4	2,880	1,440	960	720	576	480
5	5,000	2,500	1,668	1,250	1,000	834
6	7,680	3,840	2,560	1,920	1,536	1,280
7	11,040	5,520	3,860	2,760	2,205	1,840
8	15,920	7,960	5,310	3,980	3,180	2,655
10	28,800	14,400	9,600	7,200	5,750	4,800

			1/2 unit vertical in 12 units horizontal (4-percent slope)			
3	1,920	960	640	480	384	320
4	4,080	2,040	1,360	1,020	816	680
5	7,080	3,540	2,360	1,770	1,415	1,180
6	11,080	5,540	3,695	2,770	2,220	1,850
7	15,600	7,800	5,200	3,900	3,120	2,600
8	22,400	11,200	7,460	5,600	4,480	3,730
10	40,000	20,000	13,330	10,000	8,000	6,660

For SI: 1 inch = 25.4 mm, 1 square foot = 0.0929 m^2.

TABLE 11.2 Sizing chart for horizontal storm drainage piping. *Copyright 2006, International Code Council, Inc., Falls Church, Virginia. Reproduced with permission. All rights reserved.*

SIZE OF HORIZONTAL PIPING (inches)	HORIZONTALLY PROJECTED ROOF AREA (square feet)					
	Rainfall rate (inches per hour)					
	1	2	3	4	5	6
¹/₈ unit vertical in 12 units horizontal (1-percent slope)						
3	3,288	1,644	1,096	822	657	548
4	7,520	3,760	2,506	1,800	1,504	1,253
5	13,360	6,680	4,453	3,340	2,672	2,227
6	21,400	10,700	7,133	5,350	4,280	3,566
8	46,000	23,000	15,330	11,500	9,200	7,600
10	82,800	41,400	27,600	20,700	16,580	13,800
12	133,200	66,600	44,400	33,300	26,650	22,200
15	218,000	109,000	72,800	59,500	47,600	39,650
¹/₄ unit vertical in 12 units horizontal (2-percent slope)						
3	4,640	2,320	1,546	1,160	928	773
4	10,600	5,300	3,533	2,650	2,120	1,766
5	18,880	9,440	6,293	4,720	3,776	3,146
6	30,200	15,100	10,066	7,550	6,040	5,033
8	65,200	32,600	21,733	16,300	13,040	10,866
10	116,800	58,400	38,950	29,200	23,350	19,450
12	188,000	94,000	62,600	47,000	37,600	31,350
15	336,000	168,000	112,000	84,000	67,250	56,000
¹/₂ unit vertical in 12 units horizontal (4-percent slope)						
3	6,576	3,288	2,295	1,644	1,310	1,096
4	15,040	7,520	5,010	3,760	3,010	2,500
5	26,720	13,360	8,900	6,680	5,320	4,450
6	42,800	21,400	13,700	10,700	8,580	7,140
8	92,000	46,000	30,650	23,000	18,400	15,320
10	171,600	85,800	55,200	41,400	33,150	27,600
12	266,400	133,200	88,800	66,600	53,200	44,400
15	476,000	238,000	158,800	119,000	95,300	79,250

For SI: 1 inch = 25.4 mm, 1 square foot = 0.0929 m².

TABLE 11.3 Size of vertical conductors and leaders. *Copyright 2006, International Code Council, Inc., Falls Church, Virginia. Reproduced with permission. All rights reserved.*

DIAMETER OF LEADER (inches)[a]	HORIZONTALLY PROJECTED ROOF AREA (square feet)											
	Rainfall rate (inches per hour)											
	1	2	3	4	5	6	7	8	9	10	11	12
2	2,880	1,440	960	720	575	480	410	360	320	290	260	240
3	8,800	4,400	2,930	2,200	1,760	1,470	1,260	1,100	980	880	800	730
4	18,400	9,200	6,130	4,600	3,680	3,070	2,630	2,300	2,045	1,840	1,675	1,530
5	34,600	17,300	11,530	8,650	6,920	5,765	4,945	4,325	3,845	3,460	3,145	2,880
6	54,000	27,000	17,995	13,500	10,800	9,000	7,715	6,750	6,000	5,400	4,910	4,500
8	116,000	58,000	38,660	29,000	23,200	19,315	16,570	14,500	12,890	11,600	10,545	9,600

For SI: 1 inch = 25.4 mm, 1 square foot = 0.0929 m².
a. Sizes indicated are the diameter of circular piping. This table is applicable to piping of other shapes provided the cross-sectional shape fully encloses a circle of the diameter indicated in this table.

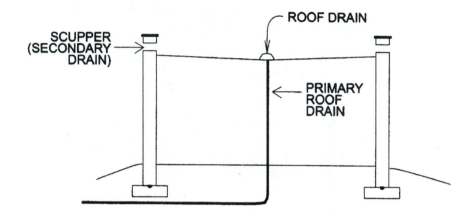

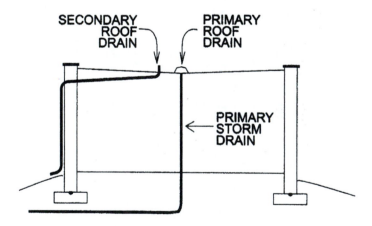

FIGURE 11.1 Separate primary and secondary roof drains. *Copyright 2002, International Code Council, Inc., Falls Church, Virginia. Reproduced with permission. All rights reserved.*

TABLE 11.4 Approved materials for building storm sewer pipe.
Copyright 2006, International Code Council, Inc., Falls Church, Virginia.
Reproduced with permission. All rights reserved.

MATERIAL	STANDARD
Acrylonitrile butadiene styrene (ABS) plastic pipe	ASTM D 2661; ASTM D 2751; ASTM F 628; CSA B181.1; CSA B182.1
Asbestos-cement pipe	ASTM C 428
Cast-iron pipe	ASTM A 74; ASTM A 888; CISPI 301
Concrete pipe	ASTM C 14; ASTM C 76; CSA A257.1M; CSA A257.2M
Copper or copper-alloy tubing (Type K, L, M or DWV)	ASTM B 75; ASTM B 88; ASTM B 251; ASTM B 306
Polyvinyl chloride (PVC) plastic pipe (Type DWV, SDR26, SDR35, SDR41, PS50 or PS100)	ASTM D 2665; ASTM D 3034; ASTM F 891; CSA B182.4; CSA B181.2; CSA B182.2
Vitrified clay pipe	ASTM C 4; ASTM C 700
Stainless steel drainage systems, Type 316L	ASME A112.3.1

TABLE 11.5 Approved materials for subsoil drain pipes in a storm water system. *Copyright 2006, International Code Council, Inc., Falls Church, Virginia. Reproduced with permission. All rights reserved.*

MATERIAL	STANDARD
Asbestos-cement pipe	ASTM C 508
Cast-iron pipe	ASTM A 74; ASTM A 888; CISPI 301
Polyethylene (PE) plastic pipe	ASTM F 405; CSA B182.1; CSA B182.6; CSA B182.8
Polyvinyl chloride (PVC) Plastic pipe (type sewer pipe, PS25, PS50 or PS100)	ASTM D 2729; ASTM F 891; CSA B182.2; CSA B182.4
Stainless steel drainage systems, Type 316L	ASME A112.3.1
Vitrified clay pipe	ASTM C 4; ASTM C 700

—NOTES—

Chapter 12
SPECIAL PIPING AND STORAGE SYSTEMS

Not all plumbers concern themselves with learning all that the code has to say about requirements for indirect-waste piping and special wastes. Some of these plumbers feel that indirect and special wastes don't affect them. However, indirect and special wastes can play a significant role in the life of a plumber. For example, the standpipes for automatic clothes washers are indirect wastes. A lot of plumbers never consider standpipe receptors as indirect wastes, but they are. The point is, whether you are doing commercial or residential work, you are likely to work with some form of indirect or special waste.

The subject of indirect and special waste is a limited one. Unlike normal drainage work, there are relatively few rules and regulations pertaining to these wastes. The rules that do apply are, however, important.

Many indirect and special waste arrangements are required because the substances being discharged are potentially harmful. This fact alone makes the topic one that every professional plumber should become well acquainted with.

Residential plumbers seem to feel the least need to know about indirect and special wastes. They don't seem to think that they will deal with such circumstances, but this just isn't the case. For example, an air gap for a dishwasher is an indirect waste. The standpipe for a laundry hookup is an indirect waste, and there can be other types of indirect or special wastes encountered by residential plumbers.

Commercial plumbers do tend to work with indirect and special wastes more than residential plumbers do, but all plumbers must learn the rules and regulations pertaining to these types of wastes. You can expect to see questions on your licensing exam that are written around indirect and special wastes. If you want to pass your test, you'd better learn all you can about them. With this in mind, let's move on to the multiple-choice exam on this subject.

MULTIPLE-CHOICE EXAM

1.   A piping arrangement in which a drain from a fixture discharges indirectly into another fixture, such as when a clothes washer discharges into a laundry tub, at a point below the flood-level rim is known as:

    a. An air gap
    b. An air break
    c. A receptor drain
    d. Both A and C

2.   Pipes that discharge into a drainage system but are not connected directly to the system are called:

    a. Direct wastes
    b. Indirect wastes
    c. Complex wastes
    d. Special wastes

3.   For a basin to be considered a swimming pool, it must have a depth of at least _____ feet at any point. Fill in the blank:

    a. 2
    b. 3
    c. 4
    d. 5

4. The unobstructed vertical distance through free atmosphere between the outlet of a waste pipe and the flood-level rim of a receptacle into which a waste pipe discharges is known as:

    a. An air break   b. An air gap

    c. A vertical gap   d. A free fall

5. Food-handling equipment must have its waste discharged through:

    a. A direct connection   b. An indirect waste pipe

    c. A grease interceptor   d. Both A and C

6. Equipment and fixtures used to store, prepare, or handle food must have their waste discharge through:

    a. A direct connection   b. An indirect waste pipe

    c. An interceptor   d. A check valve

7. An exception to the rule stated in question 6 is:

    a. A dishwashing machine   b. A dishwashing sink

    c. Both A and B   d. None of the above

8. When a device such as a sterilizer discharges potable water into a building drainage system, the discharge must be made through an indirect-waste pipe by means of:

    a. Gravity   b. An air gap

    c. An air break   d. Both A and B

9. When a device such as a process tank discharges nonpotable water into a building drainage system, the discharge must be made through an indirect-waste pipe by means of:

    a. Gravity   b. An air gap

    c. An air break   d. Both A and B

10. All indirect-waste piping must discharge through:

    a. An air gap or air break   b. A 2-inch or larger pipe

    c. A filter   d. All of the above

11. Waste receptors and standpipes must be:

    a. Trapped   b. Vented

    c. Both A and B   d. Either A or B

12. All waste receptors and standpipes must connect to:

    a. Storm-water drainage systems   b. Dry-well drainage systems

    c. Building drainage systems   d. French drains

13.   An indirect-waste pipe more than 2 feet in developed length, measured horizontally, must be:

    a.   Trapped   b.   Vented

    c.   Trapped and vented   d.   Marked as an indirect-waste pipe

14.   An indirect-waste pipe that measures more than 4 feet in developed length [measured horizontally?] must be:

    a.   Trapped   b.   Vented

    c.   Trapped and vented   d.   Marked as an indirect-waste pipe

15.   An air break is required between the indirect-waste pipe and the _____ of the waste receptor or stand-pipe. Fill in the blank:

    a.   Air gap   b.   Developed length

    c.   Trap seal   b.   Flood-level rim

16.   The air gap between an indirect-waste pipe and the flood-level rim of the waste receptor must be a minimum of _____ the effective opening of the indirect-waste pipe:

    a.   The same size as   b.   Twice

    c.   Four times   d.   None of the above

17.   Equipment utilized to store food must discharge through an indirect-waste pipe by means of:

    a.   An air gap   b.   An air break

    c.   An air hose   d.   An air relief

18.   Which of the following is required to cover the waste outlet of waste receptors:

    a.   A removable strainer   b.   A removable basket

    c.   Both A and B   d.   Either A or B

19.   Waste receptors are to be installed in:

    a.   Bathrooms   b.   Kitchens

    c.   Ventilated areas   d.   Restricted areas

20.   Waste receptors must be:

    a.   Installed in bathrooms   b.   Accessible

    c.   Readily accessible   d.   At least 3 inches in diameter

21.   Waste receptors may not be installed in:

    a.   Bathrooms   b.   Toilet rooms

    c.   Both A and B   d.   None of the above

22. Waste receptors must be installed to prevent:

    a. Streaking

    b. Splashing

    c. Osmosis

    d. Singular displacement

23. Which of the following types of waste may be discharged into the hub of a pipe:

    a. Chemical wastes

    b. Special wastes

    c. Clear-water wastes

    d. Any of the above

24. If the hub of a pipe is used as an indirect-waste receptor, the hub must extend at least _____ inches above a water-impervious floor. Fill in the blank:

    a. 2

    b. 3

    c. 4

    d. 6

25. When the hub of a pipe is used as an indirect-waste receptor, a _____ is not required. Fill in the blank with the most appropriate answer:

    a. Discharge tube

    b. Discharge relief vent

    c. Straine

    d. Receptor

26. Standpipes require _____. Fill in the blank:

    a. Strainers

    b. Baskets

    c. Individual traps

    d. Screens

27. The minimum height for a standpipe is:

    a. 12 inches

    b. 16 inches

    c. 18 inches

    d. 48 inches

28. The maximum height for a standpipe is:

    a. 31 inches

    b. 32 inches

    c. 42 inches

    d. 48 inches

29. Standpipes with a diameter of 1 1/2 inches that serve domestic automatic clothes washers must not extend more than:

    a. 18 inches

    b. 28 inches

    c. 30 inches

    d. 32 inches

30. The traps of all standpipes must be:

    a. Accessible

    b. Readily accessible

    c. Crown-vented

    d. Drum traps

31. Water being discharged into the drainage system of a building's plumbing must not exceed a temperature of:

    a. 120 degrees F.   b. 130 degrees F.

    c. 140 degrees F.   d. 160 degrees F.

32. Drains conveying water at a temperature in excess of that allowed for discharge into a drainage system directly must be piped into the drainage system by means of:

    a. A cool-down chamber   b. An indirect-waste receptor

    c. A baffled interceptor   d. None of the above

33. Which of the following are considered special wastes:

    a. Corrosive liquids   b. Spent acids

    c. Harmful chemicals   d. All of the above

34. Special wastes cannot be introduced into a plumbing system until the waste has been:

    a. Diluted   b. Monitored

    c. Screened   d. All of the above

35. Special wastes cannot be introduced into a plumbing system until the waste has been [repeat]:

    a. Neutralized   b. Manifested

    c. Screened   d. All of the above

36. Chemical waste requires:

    a. Its own drainage system   b. Its own vent system

    c. Both A and B   d. None of the above

37. Commercial dishwashing machines must be connected to the drainage system in which of the following ways:

    a. Directly

    b. Through a metallic piping material

    c. Indirectly

    d. Both A and B

38. The waste outlet for indirect waste from food-handling equipment must terminate at least _____ inches above the flood-level rim of the receptor. Fill in the blank:

    a. 2   b. 3

    c. 4   d. 6

39. The drains, overflows, and relief vents from the water-supply system for air-conditioning units must be connected through:

     a. Direct-waste connections

     b. Indirect-waste connections

     c. Either A or B

     d. Flexible connections

40. The maximum length of an indirect-waste-to-vent connection shall not exceed:

     a. 4 feet   b. 10 feet

     c. 15 feet   d. 25 feet

41. Drinking fountains _____ be installed with indirect wastes. Fill in the blank:

     a. Shall [shall, must, will always have very similar meanings...]   b. Must

     c. May   d. Will always

42. When the hub of a fitting is used as an indirect-waste receptor, the floor covering below the receptor must be:

     a. Level

     b. Pitched away from the drain

     c. Impervious to water

     d. Both B and C

43. Substances that may be harmful to the drains in a plumbing system are considered to be:

     a. Indirect wastes   b. Direct wastes

     c. Special wastes   d. Suspicious wastes

44. Chemical waste shall not be discharged into a drainage system until:

     a. It has been treated in a manner approved by the plumbing code

     b. It has been cooled to a temperature not exceeding 130 degrees F.

     c. It has been identified as a nontoxic chemical

     d. All of the above

45. Special wastes may create:

     a. Toxic fumes   b. Vertigo

     c. Eczema   d. Both B and C

You have completed the multiple-choice exam dealing with indirect and special wastes. There is still a true-false test for you to take prior to checking your answers. Before you move on to the true-false test, I want to remind you that there are differences in the plumbing codes for various parts of the country. The code in your area may provide for answers that are different from those given for the multiple-choice questions. The occasions when this occurs should be rare, but they are possible. To be absolutely sure you have answered the questions correctly, you should refer to your local code book.

I mentioned this in the beginning of this book, but I'd like to remind you that plumbing codes are amended from time to time. These changes cannot be predicted in advance and may affect the answers to the questions contained in this book. While any changes that are likely to occur should be minimal, it will be in your best interest to use this book in conjunction with a current code book. By doing this, you can be sure that the answers given to the questions on these mock tests are in compliance with current code regulations.

The compilation of questions found throughout these chapters has been derived from my personal knowledge and from reviewing code books that are current at the time this text is being written. However, I cannot guarantee that there will not be alterations to the codes by the time you take these tests. I bring this point to light to make sure that you are well prepared for your licensing exam. Now, let's move on to the true-false questions.

TRUE-FALSE EXAM

1. Cooling and air-conditioning equipment may be separated from a drainage system with the use of an air break.

True   False

2. Food equipment must be separated from a drainage system with the use of an air gap.

True   False

3. When air-gap drainage is used, the distance of the gap must not be less than 1 inch.

True   False

4. An air break may not terminate below the rim of a receptor.

True   False

5. Vents serving indirect-waste piping may be combined with vents from a sewer once they have extended 42 inches above the highest fixture's flood-level rim.

True   False

6. Indirect-waste pipes that are less than 15 feet in length are not required to be trapped.

True   False

7. Indirect-waste pipes that are less than 15 feet in length are not required to be vented.

True   False

8.   All receptors for indirect waste must be readily accessible.

True   False

9.   Automatic clothes washers used for domestic purposes may discharge into a trap that is below the floor when the appliance is situated.

True   False

10.   A trap for a clothes washer must be roughed in at least 6 inches above the floor.

True   False

11.   A trap for a clothes washer must be roughed in at a height of no more than 18 inches above the floor.

True   False

12.   Indirect-waste piping may not be installed in a storeroom.

True   False

13.   The receptor for a clothes washer may be installed in a bathroom if the washer is installed in the bathroom.

True   False

14.   Devices used as sterile equipment must be drained through an air break.

True   False

15.   Domestic dishwashers may be connected directly to a food-waste disposer without the use of an air gap.

True   False

16.   Drinking fountains must be installed with indirect-waste pipes.

True   False

17.   When piping is installed to handle chemical waste, the master plumber installing the piping must make and keep a permanent record of the location of the piping.

True   False

18.   The location of any vent used in conjunction with drainage for chemical waste must be recorded and kept permanently by the property owner.

True   False

19.   Rules and regulations for chemical wastes do not apply to small photographic darkrooms.

True   False

20.   Chemical waste may not be discharged into the ground.

True   False

We have now completed all testing on indirect and special wastes. The rules pertaining to these wastes are necessary to protect public health and safety. Don't allow yourself to take the use of indirect and special wastes lightly. Turn to the back of the chapter and check your answers against the correct answers. Whenever you find an incorrect answer in your work, consult your code book to confirm the correct answer and to learn why the answer is what it is. When you have finished grading and reviewing these tests, you may move on to the next chapter.

MULTIPLE-CHOICE ANSWERS

1.	b	10.	a	19.	c	28.	c	37.	c
2.	b	11.	c	20.	c	29.	c	38.	a
3.	a	12.	c	21.	c	30.	a	39.	b
4.	b	13.	a	22.	b	31.	c	40.	d
5.	b	14.	a	23.	c	32.	b	41.	c
6.	b	15.	c	24.	a	33.	d	42.	c
7.	c	16.	b	25.	c	34.	a	43.	c
8.	b	17.	a	26.	c	35.	a	44.	a
9.	c	18.	d	27.	c	36.	c	45.	a

TRUE-FALSE ANSWERS

1.	True	6.	False	11.	True	16.	False
2.	True	7.	True	12.	True	17.	False
3.	True	8.	True	13.	True	18.	True
4.	False	9.	False	14.	False	19.	True
5.	False	10.	True	15.	False	20.	True

Chapter 13
RECYCLING GRAY WATER

The recycling of gray water is not common in many areas, but it is very useful in some situations. When a recycling system is designed and installed, it can take load off a sewer. Regions where water is scarce can benefit greatly from the recycling of gray water. While you may not be called upon to work with a gray-water recycling system, you might see some questions about the procedure on your licensing exam. Since this is a short topic, we will be dealing only with true-false questions in this chapter.

TRUE-FALSE EXAM

1. Any residential fixture can discharge into a gray-water system.

   True   False

2. Bathtubs and showers may discharge into a gray-water system.

   True   False

3. Only lavatories can discharge into a gray-water system.

   True   False

4. Lavatories and clothes washers may discharge into a gray-water system.

   True   False

5. Laundry sinks may discharge into a gray-water system.

   True   False

6. Recycled gray water can be reintroduced into a water heater.

   True   False

7. Gray water can be used to flush toilets and urinals located in the same building as the gray-water recycling system.

   True   False

8. When approved, gray water can be used for irrigation purposes.

   True   False

9. Gray-water reservoirs must be constructed of concrete.

   True   False

10. All gray-water reservoirs are required to be closed and gas-tight.

True   False

11. Access openings must be provided to allow inspection and cleaning of a gray-water reservoir.

True   False

12. The minimum capacity of a gray-water reservoir must be four times the volume of water required to meet the daily flushing requirements of the fixtures supplied with gray water but not less than 50 gallons.

True   False

13. A gray-water reservoir must be sized to limit the retention time of gray water to a maximum of 72 hours.

True   False

14. Gray water must be filtered as it enters a reservoir.

True   False

15. A media, sand, or diatomaceous-earth filter may be used to filter gray water entering a reservoir.

True   False

16. Gray water must be disinfected.

True   False

17. Chlorine may be used to disinfect gray water.

True   False

18. Potable water may not be used as a source of makeup water for a gray-water system.

True   False

19. When a makeup water-supply line is connected to a gray-water system, the supply line must be equipped with a full-open valve.

True   False

20. Overflow pipes are not required for gray-water systems.

True   False

21. Any overflow pipe in a gray-water system must not be connected to a sanitary drainage system.

True   False

22.   Gray-water reservoirs must be vented.
      True   False

23.   Gray water must be dyed blue or green with a food-grade vegetable dye before the water is supplied to fixtures.
      True   False

24.   Gray-water piping must be identified as containing nonpotable water.
      True   False

25.   A drain is required at the lowest point of a gray-water collection reservoir.
      True   False

26.   Diatomaceous earth is ok for use in collection reservoirs?
      True   False

27.   A shower is an approved fixture that is allowed to discharge gray water.
      True   False

28.   A collection area must be left open for easy access.
      True   False

29.   Gray water must be disinfected with an approved substance, such as alcohol.
      True   False

30.   Identification of all distribution pipes must be approved by local code authority as a gray-water system.
      True   False

31.   Site locations for a subsurface landscape irrigation system is regulated.
      True   False

32.   When constructing a seepage bed or trench, the bottom must be uneven.
      True   False

33.   A 2-inch cover of aggregate is required over the top of distribution piping.
      True   False

34.   Building paper is an approved matter as a cover for the aggregate.
      True   False

35.   When installed, distribution piping must have a minimum of eight inches of soil cover.
          True   False

36.   In no case is the holding capacity of a reservoir to be less than 50 gallons.
          True   False

37. $D = A \times B$ is the equation that is used to calculate the gray water discharge.
          True   False

38.   Recycling gray water is good for the environment and it can save money in the long run.
          True   False

39.   A collection area must be closed and gas tight.
          True   False

40.   An access opening is required so that interior inspections can be performed.
          True   False

FILL IN THE BLANK EXAM

1.   All gray water going into a collection reservoir must be _____.

2.   Retention time for gray water is _____ hours?

3.   Gray water must be dyed blue or green with a food-grade _____ dye.

4.   A _____ test is required to determine the perk rate of the soil.

5.   _____ irrigation systems that use gray water must be installed at an elevation which is lower than the surface grade.

6.   Seepage trenches must have a minimum width of _____ foot and a maximum width of _____ feet.

7.   Trenches are to be spaced a minimum of _____ feet apart.

8.   Seepage beds are required to have more than one _____ pipe.

9.   _____ must meet local requirements and may have to be sacrificed.

10.   _____ is required to have a minimum depth of nine inches.

Cross off the answers as you use them

one	filtered	Sidewalls	perk	72	2	five

vegetable	Soil cover	distribution	Subsurface

TRUE-FALSE ANSWERS

1. False	11. True	21. False	31. True
2. True	12. False	22. True	32. False
3. False	13. True	23. True	33. True
4. True	14. True	24. True	34. False
5. True	15. True	25. True	35. True
6. False	16. True	26. True	36. True
7. True	17. True	27. True	37. False
8. True	18. False	28. False	38. True
9. False	19. True	29. False	39. True
10. True	20. False	30. True	40. True

FILL IN THE BLANK ANSWERS

1. filtered	6. 1 and 5
2. 72	7. 2
3. vegetable	8. distribution
4. perk	9. Sidewalls
5. Subsurface	10. Soil cover

Chapter 14

REFERENCED STANDARDS

ANSI

American National Standards Institute
25 West 43rd Street, Fourth Floor
New York, NY 10036

Standard Reference Number	Title	Referenced in code section number
Z4.3—95	Minimum Requirements for Nonsewered Waste-Disposal Systems	311.1
Z21.22—99 (R2003)	Relief Valves for Hot Water Supply Systems with Addenda Z21.22a-2000(R2003) and Z21.22b-2001(R2003)	504.2, 504.5
Z124.1—95	Plastic Bathtub Units	407.1
Z124.2—95	Plastic Shower Receptors and Shower Stalls	417.1
Z124.3—95	Plastic Lavatories	416.1, 416.2
Z124.4—96	Plastic Water Closet Bowls and Tanks	420.1
Z124.6—97	Plastic Sinks	415.1, 418.1
Z124.9-94	Plastic Urinal Fixtures	419.1

ARI

Air-Conditioning & Refrigeration Institute
4100 North Fairfax Drive, Suite 200
Arlington, VA 22203

Standard reference number	Title	Referenced in code section number
1010—02	Self-contained, Mechanically Refrigerated Drinking-water Coolers	410.1

ASME

American Society of Mechanical Engineers
Three Park Avenue
New York, NY 10016-5990

Standard Reference Number	Title	Referenced in code section number
A112.1.2—1991(R2002)	Air Gaps in Plumbing Systems	Table 608.1
A112.1.3—2000	Air Gap Fittings for Use with Plumbing Fixtures, Appliances and Appurtenances	608.13.1, Table 608.1
A112.3.1—1993	Performance Standard and Installation Procedures for Stainless Steel Drainage Systems for Sanitary, Storm and Chemical Applications, Above and Below Ground	412.1, Table 702.1, Table 702.2, Table 702.3, Table 702.4, 708.2, Table 1102.4, Table 1102.5, 1102.6, Table 1102.7
A112.3.4—2000	Macerating Toilet Systems and Related Components	712.4.1
A112.4.1—1993(R2002)	Water Heater Relief Valve Drain Tubes	504.6.2
A112.4.3—1999	Plastic Fittings for Connecting Water Closets to the Sanitary Drainage System	405.4
A112.6.1M—1997(R2002)	Floor-Affixed Supports for Off-the-floor Plumbing Fixtures for Public Use	405.4.3
A112.6.2—2000	Framing-Affixed Supports for Off-the-floor Water Closets with Concealed Tanks	405.4.3
A112.6.3—2001	Floor and Trench Drains	412.1
A112.6.7—2001	Enameled and Epoxy-coated Cast-iron and PVC Plastic Sanitary Floor Sinks	427.1
A112.14.1—2003	Backwater Valves	715.2
A112.14.3—2000	Grease Interceptors	1003.3.4
A112.14.4—2001	Grease Removal Devices	1003.3.4
A112.18.1—2003	Plumbing Fixture Fittings	424.1, 608.2
A112.18.2—2002	Plumbing Fixture Waste Fittings	424.1.2
A112.18.3—2002	Performance Requirements for Backflow Protection Devices and Systems in Plumbing Fixture Fittings	424.4
A112.18.6—2003	Flexible Water Connectors	605.6
A112.18.7—1999	Deck mounted Bath/Shower Transfer Valves with Integral Backflow Protection	424.6
A112.19.1M—1994(R1999)	Enameled Cast Iron Plumbing Fixtures with 1998 and 2000 supplements	407.1, 410.1, 415.1, 416.1, 418.1

ASSE

American Society of Sanitary Engineering
901 Canterbury Road, Suite A
Westlake, OH 44145

Standard Reference Number	Title	Referenced in code section number
1001—02	Performance Requirements for Atmospheric Type Vacuum Breakers	425.2, Table 608.1, 608.13.6
1002—99	Performance Requirements for Antisiphon Fill Valves (Ballcocks) for Gravity Water Closet Flush Tanks	425.3.1, Table 608.1
1003—01	Performance Requirements for Water Pressure Reducing Valves	604.8
1004—90	Performance Requirements for Backflow Prevention Requirements for Commercial Dishwashing Machines	409.1
1005—99	Performance Requirements for Water Heater Drain Valves	501.3
1006—89	Performance Requirements for Residential Use Dishwashers	409.1
1007—92	Performance Requirements for Home Laundry Equipment	406.1, 406.2
1008—89	Performance Requirements for Household Food Waste Disposer Units	413.1
1009—90	Performance Requirements for Commercial Food Waste Grinder Units	413.1
1010—96	Performance Requirements for Water Hammer Arresters	604.9
1011—93	Performance Requirements for Hose Connection Vacuum Breakers	Table 608.1, 608.13.6
1012—02	Performance Requirements for Backflow Preventers with Intermediate Atmospheric Vent	Table 608.1, 608.13.3, 608.16.2
1013—99	Performance Requirements for Reduced Pressure Principle Backflow Preventers and Reduced Pressure Fire Protection Principle Backflow Preventers	Table 608.1, 608.13.2, 608.16.2
1014—90	Performance Requirements for Handheld Showers	424.2
1015—99	Performance Requirements for Double Check Backflow Prevention Assemblies and Double Check Fire Protection Backflow Prevention Assemblies	Table 608.1, 608.13.7
1016—96	Performance Requirements for Individual Thermostatic, Pressure Balancing and Combination Control Valves for Individual Fixture Fittings	424.3
1017—99	Performance Requirements for Temperature Actuated Mixing Valves for Hot Water Distribution Systems	424.3, 424.5, 613.1
1018—01	Performance Requirements for Trap Seal Primer Valves; Potable Water Supplied	1002.4
1019—97	Performance Requirements for Vacuum Breaker Wall Hydrants, Freeze Resistant, Automatic Draining Type	Table 608.1, 608.13.6

ASSE—continued

5048—98	Performance Requirements for Testing Double Check Valve Detector Assembly (DCDA).	312.9.2
5052—98	Performance Requirements for Testing Hose Connection Backflow Preventers.	312.9.2
5056—98	Performance Requirements for Testing Spill Resistant Vacuum Breaker	312.9.2

ASTM

ASTM International
100 Barr Harbor Drive
West Conshohocken, PA 19428-2959

Standard Reference Number	Title	Referenced in code section number
A 53/A 53M—02	Specification for Pipe, Steel, Black and Hot-dipped, Zinc-coated Welded and Seamless.	Table 605.3, Table 605.4, Table 702.1
A 74—04a	Specification for Cast-iron Soil Pipe and Fittings	Table 702.1, Table 702.2, Table 702.3, Table 702.4, 708.2, Table 1102.5, Table 1102.7
A 312/A 312M—04a	Specification for Seamless and Welded Austenitic Stainless Steel Pipes	Table 605.4, Table 605.5, Table 605.6, 605.23.2
A 733—03	Specification for Welded and Seamless Carbon Steel and Austenitic Stainless Steel Pipe Nipples	Table 605.8
A 778—01	Specification for Welded Unannealed Austenitic Stainless Steel Tubular Products	Table 605.4, Table 605.5, Table 605.6
A 888—04a	Specification for Hubless Cast-iron Soil Pipe and Fittings for Sanitary and Storm Drain, Waste, and Vent Piping Application.	Table 702.1, Table 702.2, Table 702.3, Table 702.4, Table 1102.4, Table 1102.5, Table 1102.7
B 32—03	Specification for Solder Metal.	605.14.3, 605.15.4, 705.9.3, 705.10.3
B 42—02e01	Specification for Seamless Copper Pipe, Standard Sizes	Table 605.3, Table 605.4, Table 702.1
B 43—(2004)	Specification for Seamless Red Brass Pipe, Standard Sizes.	Table 605.3, Table 605.4, Table 702.1
B 75—02	Specification for Seamless Copper Tube.	Table 605.3, Table 605.4, Table 702.1, Table 702.2, Table 702.3, Table 605.4, Table 1102.4
B 88—03	Specification for Seamless Copper Water Tube.	Table 605.3, Table 605.4, Table 702.1, Table 702.2, Table 702.3, Table 1102.4
B 152/B 152M—00	Specification for Copper Sheet, Strip Plate and Rolled Bar.	402.3, 425.3.3, 417.5.2.4, 902.2
B 251—02e01	Specification for General Requirements for Wrought Seamless Copper and Copper-alloy Tube.	Table 605.3, Table 605.4, Table 702.1, Table 702.2, Table 702.3, Table 1102.4

ASTM—continued

ASTM—continued

ASTM—continued

Waste, and Vent Pipe with a Cellular Core .... Table 702.1, Table 702.2, Table 702.3, Table 702.4, 705.2.2, 705.7.2, Table 1102.4, Table 1102.7

F 656—02 Specification for Primers for Use in Solvent Cement Joints of Poly (Vinyl Chloride) (PVC) Plastic Pipe and Fittings 605.21.2, 705.8.2, 705.14.2

F 714—03 Specification for Polyethylene (PE) Plastic Pipe (SDR-PR) Based on Outside Diameter .... Table 702.3

F 876—04 Specification for Cross-linked Polyethylene (PEX) Tubing .... Table 605.3

F 877—02e01 Specification for Cross-linked Polyethylene (PEX) Plastic Hot and Cold Water Distribution Systems .... Table 605.3, Table 605.4, Table 605.5, Table 605.17.2

F 891—00e01 Specification for Coextruded Poly (Vinyl Chloride) (PVC) Plastic Pipe with a Cellular Core .... Table 702.1, Table 702.2, Table 702.3, Table 702.4, Table 1102.4, Table 1102.5, Table 1102.7

F 1281—03 Specification for Cross-linked Polyethylene/Aluminum/Cross-Linked Polyethylene (PEX-AL-PEX) Pressure Pipe .... Table 605.3, Table 605.4

F 1282—03 Specification for Polyethylene/Aluminum/Polyethylene (PE-AL-PE) Composite Pressure Pipe .... Table 605.3, Table 605.4

F 1412—01 Specification for Polyolefin Pipe and Fittings for Corrosive Waste Drainage .... Table 702.2, Table 702.4, 705.17.1

F 1488—03 Specification for Coextruded Composite Pipe .... Table 702.1, Table 702.2, Table 702.3

F 1807—04 Specification for Metal Insert Fittings Utilizing a Copper Crimp Ring for SDR9 Cross-linked Polyethylene (PEX) Tubing .... Table 605.5, 605.17.2

F 1866—98 Specification for Poly (Vinyl Chloride) (PVC) Plastic Schedule 40 Drainage and DWV Fabricated Fittings .... Table 702.4

F 1960—04a Specification for Cold Expansion Fittings with PEX Reinforcing Rings for use with Cross-linked Polyethylene (PEX) Tubing .... Table 605.5

F 1974—04 Specification for Metal Insert Fittings for Polyethylene/Aluminum/Polyethylene and Cross-linked Polyethylene/Aluminum/Cross-linked Polyethylene Composite Pressure Pipe .... Table 605.5

F 1986—00a Specification for Multilayer Pipe, Type 2, Compression Fittings and Compression Joints for Hot and Cold Drinking Water Systems .... Table 605.3, Table 605.4, Table 605.5

F 2080—04 Specifications for Cold-expansion Fittings with Metal Compression-sleeves for Cross-linked Polyethylene (PEX) Pipe .... Table 605.5

F 2159—03 Specification for Plastic Insert Fittings Utilizing a Copper Crimp Ring for SDR9 Cross-linked Polyethylene (PEX) Tubing .... Table 605.5

F 2389—04 Specification for Pressure-rated Polypropylene (PP) Piping Systems .... Table 605.3, Table 605.4, Table 605.5, 605.21

AWS

American Welding Society
550 N.W. LeJeune Road
Miami, FL 33126

Standard Reference Number	Title	Referenced in code section number
A5.8—04	Specifications for Filler Metals for Brazing and Braze Welding	605.12.1, 605.14.1, 605.15.1, 705.4.1, 705.9.1, 705.10.1

AWWA

American Water Works Association
6666 West Quincy Avenue
Denver, CO 80235

Standard Reference Number	Title	Referenced in code section number
C104—98	Standard for Cement-mortar Lining for Ductile-Iron Pipe and Fittings for Water	605.3, 605.5
C110—98	Standard for Ductile-iron and Gray-iron Fittings, 3 Inches through 48 Inches, for Water	Table 605.5, Table 702.4, Table 1102.7
C111—00	Standard for Rubber-gasket Joints for Ductile-iron Pressure Pipe and Fittings	605.13
C115—99	Standard for Flanged Ductile-iron Pipe with Ductile-iron or Gray-iron Threaded Flanges	Table 605.3
C151/A21.51—02	Standard for Ductile-iron Pipe, Centrifugally Cast for Water	Table 605.3
C153—00	Standard for Ductile-iron Compact Fittings for Water Service	Table 605.5
C510—00	Double Check Valve Backflow Prevention Assembly	Table 608.1, 608.13.7
C511—00	Reduced-pressure Principle Backflow Prevention Assembly	Table 608.1, 608.13.2, 608.16.2
C651—99	Disinfecting Water Mains	610.1
C652—02	Disinfection of Water-storage Facilities	610.1

CISPI

Cast Iron Soil Pipe Institute
5959 Shallowford Road, Suite 419
Chattanooga, TN 37421

Standard Reference Number	Title	Referenced in code section number
301—04a	Specification for Hubless Cast-iron Soil Pipe and Fittings for Sanitary and Storm Drain, Waste and Vent Piping Applications	Table 702.1, Table 702.2, Table 702.3, Table 702.4, Table 1102.4, Table 1102.5, Table 1102.7
310—04	Specification for Coupling for Use in Connection with Hubless Cast-iron Soil Pipe and Fittings for Sanitary and Storm Drain, Waste and Vent Piping Applications	705.5.3

CSA

Canadian Standards Association
178 Rexdale Blvd.
Rexdale (Toronto), Ontario, Canada M9W 1R3

Standard Reference Number	Title	Referenced in code section number
B45.1—02	Ceramic Plumbing Fixtures	408.1, 416.1, 418.1, 419.1, 420.1
B45.2—02	Enameled Cast-iron Plumbing Fixtures	407.1, 415.1, 416.1, 418.1
B45.3—02	Porcelain Enameled Steel Plumbing Fixtures	407.1, 416.1, 418.1
B45.4—02	Stainless-steel Plumbing Fixtures	415.1, 416.1, 418.1, 420.1
B45.5—02	Plastic Plumbing Fixtures	407.1, 416.2, 417.1, 419.1, 420.1, 421.1
B45.9—99	Macerating Systems and Related Components	712.4.1
B45.10—01	Hydromassage Bathtubs	421.1
B64.1.2—01	Vacuum Breakers, Pressure Type (PVB)	Table 608.1, 608.13.5
B64.2.1—01	Vacuum Breakers, Hose Connection Type (HCVB) with Manual Draining Feature	Table 608.1, 608.13.6
B64.2.1.1—01	Vacuum Breakers, Hose Connection Dual Check Type (HCDVB)	Table 608.1, 608.13.6
B64.3.1—01	Backflow Preventers, Dual Check Valve Type with Atmospheric Port for Carbonators (DCAPC)	Table 608.1, 608.16.1
B64.4.1—01	Backflow Preventers, Reduced Pressure Principle Type for Fire Sprinklers (RPF)	Table 608.1, 608.13.2
B64.5—01	Backflow Preventers, Double Check Type (DCVA)	Table 608.1, 608.13.7
B64.5.1—01	Backflow Preventers, Double Check Type for Fire Systems (DCVAF)	Table 608.1 608.13.7
B64.6—01	Backflow Preventers, Dual Check Valve Type (DuC)	605.3.1, Table 608.1

CSA—continued

Standard Reference Number	Title	Referenced in code section number
CAN/CSA-B64.3—01	Backflow Preventers, Dual Check Valve Type with Atmospheric Port (DCAP)	Table 608.1, 608.13.3, 608.16.2
CAN/CSA-B64.4—01	Backflow Preventers, Reduced Pressure Principle Type (RP)	Table 608.1, 608.13.2, 608.16.2
CAN/CSA-B64.10—01	Manual for the Selection, Installation, Maintenance and Field Testing of Backflow Prevention Devices	312.9.2
CAN/CSA-B137.9—02	Polyethylene/Aluminum/Polyethylene Composite Pressure Pipe Systems	Table 605.3
CAN/CSA-B137.10M—02	Cross-linked Polyethylene/Aluminum/Polyethylene Composite Pressure Pipe Systems	Table 605.3, Table 605.4
CAN/CSA-B181.3—02	Polyolefin Laboratory Drainage Systems	Table 702.1, Table 702.2
CAN/CSA-B182.4—02	Profile PVC Sewer Pipe and Fittings	Table 702.3, Table 1102.4, Table 1102.5
CAN/CSA-B602—02	Mechanical Couplings for Drain, Waste and Vent Pipe and Sewer Pipe	705.2.1, 705.5.3, 705.6, 705.7.1, 705.14.1, 705.15, 705.16

ICC

International Code Council
5203 Leesburg Pike, Suite 600
Falls Church, VA 22041

Standard Reference Number	Title	Referenced in code section number
IBC—06	International Building Code®	201.3, 305.4, 307.1, 307.2, 307.3, 308.2, 309.1, 310.1, 310.3, 403.1, Table 403.1, 404.1, 407.3, 417.6, 502.6, 606.5.2, 1106.5
ICC EC—06	ICC Electrical Code®	201.3, 502.1, 504.3, 1113.1.3
IEBC—06	International Existing Building Code®	101.2
IECC—06	International Energy Conservation Code®	313.1, 607.2, 607.2.1
IFC—06	International Fire Code®	201.3, 1201.1
IFGC—06	International Fuel Gas Code®	101.2, 201.3, 502.1
IMC—06	International Mechanical Code®	201.3, 307.6, 310.1, 422.9, 502.1, 612.1, 1202.1
IPSDC—06	International Private Sewage Disposal Code®	701.2
IRC—06	International Residential Code®	101.2

ISEA

Industry Safety Equipment Association
1901 N. Moore Street, Suite 808
Arlington, VA 22209

Standard Reference Number	Title	Referenced in code section number
Z358.1—03	Emergency Eyewash and Shower Equipment.	411.1

NFPA

National Fire Protection Association
Batterymarch Park
Quincy, MA 02269

Standard Reference Number	Title	Referenced in code section number
50—01	Bulk Oxygen Systems at Consumer Sites.	1203.1
51—02	Design and Installation of Oxygen-fuel Gas Systems for Welding, Cutting and Allied Processes	1203.1
70—05	National Electrical Code.	502.1, 504.3, 1113.1.3
99C—02	Gas and Vacuum Systems	1202.1

NSF

NSF International
789 Dixboro Road
Ann Arbor, MI 48105

Standard Reference Number	Title	Referenced in code section number
3—2003	Commercial Warewashing Equipment	409.1
14—2003	Plastic Piping System Components and Related Materials	303.3, 611.3
18—2004	Manual Food and Beverage Dispensing Equipment	426.1
42—2002e	Drinking Water Treatment Units—Aesthetic Effects	611.1, 611.3
44—2004	Residential Cation Exchange Water Softeners	611.1, 611.3
53—2002e	Drinking Water Treatment Units—Health Effects	611.1, 611.3
58—2004	Reverse Osmosis Drinking Water Treatment Systems	611.2
61—2003e	Drinking Water System Components—Health Effects	424.1, 605.3, 605.4, 605.5, 611.3
62—2004	Drinking Water Distillation Systems	611.1

PDI

Plumbing and Drainage Institute
800 Turnpike Street, Suite 300
North Andover, MA 01845

Standard Reference Number	Title	Referenced in code section number
G101 (2003)	Testing and Rating Procedure for Grease Interceptors with Appendix of Sizing and Installation Data	1003.3.4

UL

Underwriters Laboratories, Inc.
333 Pfingsten Road
Northbrook, IL 60062-2096

Standard reference number	Title	Referenced in code section number
UL508—99	Industrial Control Equipment	314.2.3

—NOTES—

Chapter 15
RAINFALL RATES

RATES OF RAINFALL FOR VARIOUS CITIES

Rainfall rates, in inches per hour, are based on a storm of one-hour duration and a 100-year return period.

Alabama:	**District of Columbia:**	Topeka 3.7	Biloxi 4.7
Birmingham 3.8	Washington 3.2	Wichita 3.7	Columbus 3.9
Huntsville 3.6			Corinth 3.6
Mobile 4.6	**Florida:**	**Kentucky:**	Natchez 4.4
Montgomery 4.2	Jacksonville 4.3	Ashland 3.0	Vicksburg 4.1
	Key West 4.3	Lexington 3.1	
Alaska:	Miami 4.7	Louisville 3.2	**Missouri:**
Fairbanks 1.0	Pensacola 4.6	Middlesboro 3.2	Columbia 3.2
Juneau 0.6	Tampa 4.5	Paducah 3.3	Kansas City 3.6
			Springfield 3.4
Arizona:	**Georgia:**	**Louisiana:**	St. Louis 3.2
Flagstaff 2.4	Atlanta 3.7	Alexandria 4.2	
Nogales 3.1	Dalton 3.4	Lake Providence 4.0	**Montana:**
Phoenix 2.5	Macon 3.9	New Orleans 4.8	Ekalaka 2.5
Yuma 1.6	Savannah 4.3	Shreveport 3.9	Havre 1.6
	Thomasville 4.3		Helena 1.5
Arkansas:		**Maine:**	Kalispell 1.2
Fort Smith 3.6	**Hawaii:**	Bangor 2.2	Missoula 1.3
Little Rock 3.7	Hilo 6.2	Houlton 2.1	
Texarkana 3.8	Honolulu 3.0	Portland 2.4	**Nebraska:**
	Wailuku 3.0		North Platte 3.3
California:		**Maryland:**	Omaha 3.8
Barstow 1.4	**Idaho:**	Baltimore 3.2	Scottsbluff 3.1
Crescent City 1.5	Boise 0.9	Hagerstown 2.8	Valentine 3.2
Fresno 1.1	Lewiston 1.1	Oakland 2.7	
Los Angeles 2.1	Pocatello 1.2	Salisbury 3.1	**Nevada:**
Needles 1.6			Elko 1.0
Placerville 1.5	**Illinois:**	**Massachusetts:**	Ely 1.1
San Fernando 2.3	Cairo 3.3	Boston 2.5	Las Vegas 1.4
San Francisco 1.5	Chicago 3.0	Pittsfield 2.8	Reno 1.1
Yreka 1.4	Peoria 3.3	Worcester 2.7	
	Rockford 3.2		**New Hampshire:**
Colorado:	Springfield 3.3	**Michigan:**	Berlin 2.5
Craig 1.5		Alpena 2.5	Concord 2.5
Denver 2.4	**Indiana:**	Detroit 2.7	Keene 2.4
Durango 1.8	Evansville 3.2	Grand Rapids 2.6	
Grand Junction 1.7	Fort Wayne 2.9	Lansing 2.8	**New Jersey:**
Lamar 3.0	Indianapolis 3.1	Marquette 2.4	Atlantic City 2.9
Pueblo 2.5		Sault Ste. Marie 2.2	Newark 3.1
	Iowa:		Trenton 3.1
Connecticut:	Davenport 3.3	**Minnesota:**	
Hartford 2.7	Des Moines 3.4	Duluth 2.8	**New Mexico:**
New Haven 2.8	Dubuque 3.3	Grand Marais 2.3	Albuquerque 2.0
Putnam 2.6	Sioux City 3.6	Minneapolis 3.1	Hobbs 3.0
		Moorhead 3.2	Raton 2.5
Delaware:	**Kansas:**	Worthington 3.5	Roswell 2.6
Georgetown 3.0	Atwood 3.3		Silver City 1.9
Wilmington 3.1	Dodge City 3.3	**Mississippi:**	

(continued)

FIGURE 15.1   Rates of rainfall for various cities. *Source: National Weather Service, National Oceanic and Atmospheric Administration, Washington, D.C.*

New York:
Albany 2.5
Binghamton 2.3
Buffalo 2.3
Kingston 2.7
New York 3.0
Rochester 2.2

North Carolina:
Asheville 4.1
Charlotte 3.7
Greensboro 3.4
Wilmington 4.2

North Dakota:
Bismarck 2.8
Devils Lake 2.9
Fargo 3.1
Williston 2.6

Ohio:
Cincinnati 2.9
Cleveland 2.6
Columbus 2.8
Toledo 2.8

Oklahoma:
Altus 3.7
Boise City 3.3
Durant 3.8
Oklahoma City 3.8

Oregon:
Baker 0.9
Coos Bay 1.5
Eugene 1.3
Portland 1.2

Pennsylvania:
Erie 2.6
Harrisburg 2.8
Philadelphia 3.1
Pittsburgh 2.6
Scranton 2.7

Rhode Island:
Block Island 2.75
Providence 2.6

South Carolina:
Charleston 4.3
Columbia 4.0
Greenville 4.1

South Dakota:
Buffalo 2.8
Huron 3.3
Pierre 3.1
Rapid City 2.9
Yankton 3.6

Tennessee:
Chattanooga 0.5
Knoxville 3.2
Memphis 3.7
Nashville 3.3

Texas:
Abilene 3.6
Amarillo 3.5
Brownsville 4.5
Dallas 4.0
Del Rio 4.0
El Paso 2.3
Houston 4.6
Lubbock 3.3

Odessa 3.2
Pecos 3.0
San Antonio 4.2

Utah:
Brigham City 1.2
Roosevelt 1.3
Salt Lake City 1.3
St. George 1.7

Vermont:
Barre 2.3
Bratteboro 2.7
Burlington 2.1
Rutland 2.5

Virginia:
Bristol 2.7
Charlottesville 2.8
Lynchburg 3.2
Norfolk 3.4
Richmond 3.3

Washington:
Omak 1.1
Port Angeles 1.1
Seattle 1.4
Spokane 1.0
Yakima 1.1

West Virginia:
Charleston 2.8
Morgantown 2.7

Wisconsin:
Ashland 2.5
Eau Claire 2.9
Green Bay 2.6

La Crosse 3.1
Madison 3.0
Milwaukee 3.0

Wyoming:
Cheyenne 2.2
Fort Bridger 1.3
Lander 1.5
New Castle 2.5
Sheridan 1.7
Yellowstone Park 1.4

FIGURE 15.1 *(Continued)* Rates of rainfall for various cities. *Source: National Weather Service, National Oceanic and Atmospheric Administration, Washington, D.C.*

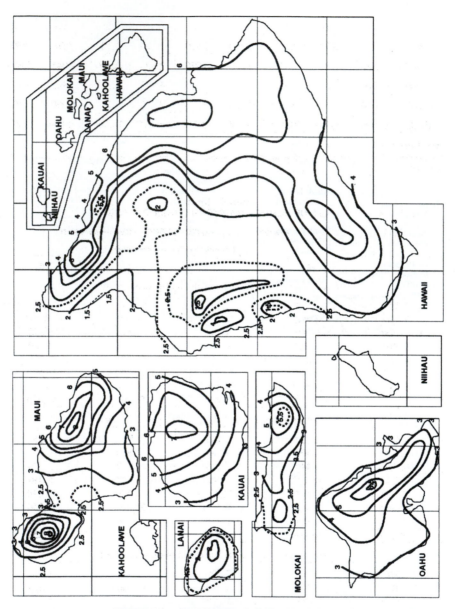

100-YEAR, 1-HOUR RAINFALL (INCHES) HAWAII

For SI: 1 inch = 25.4 mm.

FIGURE 15.2   A chart for the one hundred year, one-hour rainfall (inches) for Hawaii.
Source: National Weather Service, National Oceanic and Atmospheric Administration, Washington, D.C.

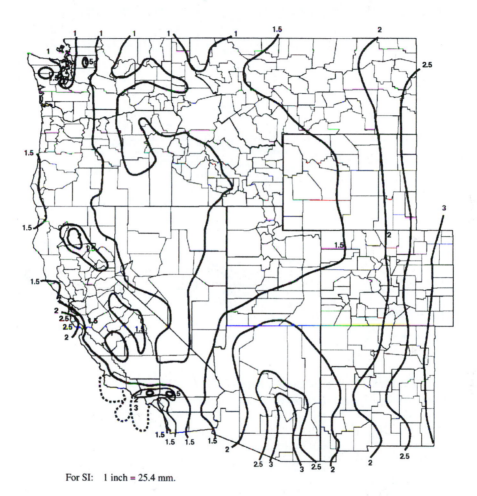

For SI: 1 inch = 25.4 mm.

FIGURE 15.3 A chart for the one hundred year, one-hour rainfall (inches) for the western United States. *Source: National Weather Service, National Oceanic and Atmospheric Administration, Washington, D.C.*

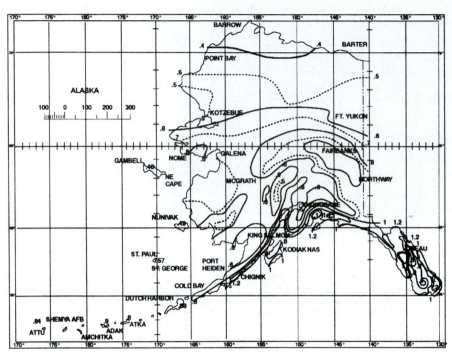

For SI: 1 inch = 25.4 mm.

FIGURE 15.4   A chart for the one hundred year, one-hour rainfall (inches) for Alaska.
Source: National Weather Service, National Oceanic and Atmospheric Administration, Washington, D.C.

For SI: 1 inch = 25.4 mm.

FIGURE 15.5 A chart for the one hundred year, one-hour rainfall (inches) for the eastern United States. *Source: National Weather Service, National Oceanic and Atmospheric Administration, Washington, D.C.*

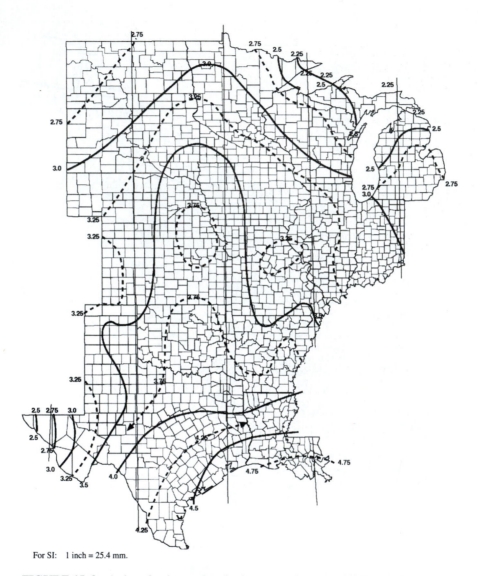

For SI: 1 inch = 25.4 mm.

FIGURE 15.6   A chart for the one hundred year, one-hour rainfall (inches) for the central United States. *Source: National Weather Service, National Oceanic and Atmospheric Administration, Washington, D.C.*

Chapter 16

DEGREE DAYS AND DESIGN TEMPERATURES

DEGREE DAY AND DESIGN TEMPERATURES[a] FOR CITIES IN THE UNITED STATES

| STATE | STATION[b] | HEATING DEGREE DAYS (yearly total) | DESIGN TEMPERATURES | | | DEGREES NORTH LATITUDE[c] |
| | | | Winter | Summer | | |
			97½%	Dry bulb 2½%	Wet bulb 2½%	
AL	Birmingham	2,551	21	94	77	33°30'
	Huntsville	3,070	16	96	77	34°40'
	Mobile	1,560	29	93	79	30°40'
	Montgomery	2,291	25	95	79	32°20'
AK	Anchorage	10,864	-18	68	59	61°10'
	Fairbanks	14,279	-47	78	62	64°50'
	Juneau	9,075	1	70	59	58°20'
	Nome	14,171	-27	62	56	64°30'
AZ	Flagstaff	7,152	4	82	60	35°10'
	Phoenix	1,765	34	107	75	33°30'
	Tuscon	1,800	32	102	71	33°10'
	Yuma	974	39	109	78	32°40'
AR	Fort Smith	3,292	17	98	79	35°20'
	Little Rock	3,219	20	96	79	34°40'
	Texarkana	2,533	23	96	79	33°30'
CA	Fresno	2,611	30	100	71	36°50'
	Long Beach	1,803	43	80	69	33°50'
	Los Angeles	2,061	43	80	69	34°00'
	Los Angeles[d]	1,349	40	89	71	34°00'
	Oakland	2,870	36	80	64	37°40'
	Sacramento	2,502	32	98	71	38°30'
	San Diego	1,458	44	80	70	32°40'
	San Francisco	3,015	38	77	64	37°40'
	San Francisco[d]	3,001	40	71	62	37°50'
CO	Alamosa	8,529	-16	82	61	37°30'
	Colorado Springs	6,423	2	88	62	38°50'
	Denver	6,283	1	91	63	39°50'
	Grand Junction	5,641	7	94	63	39°10'
	Pueblo	5,462	0	95	66	38°20'
CT	Bridgeport	5,617	9	84	74	41°10'
	Hartford	6,235	7	88	75	41°50'
	New Haven	5,897	7	84	75	41°20'
DE	Wilmington	4,930	14	89	76	39°40'
DC	Washington	4,224	17	91	77	38°50'
FL	Daytona	879	35	90	79	29°10'
	Fort Myers	442	44	92	79	26°40'
	Jacksonville	1,239	32	94	79	30°30'
	Key West	108	57	90	79	24°30'
	Miami	214	47	90	79	25°50'
	Orlando	766	38	93	78	28°30'
	Pensacola	1,463	29	93	79	30°30'
	Tallahassee	1,485	30	92	78	30°20'
	Tampa	683	40	91	79	28°00'
	West Palm Beach	253	45	91	79	26°40'

DEGREE DAY AND DESIGN TEMPERATURES[a] FOR CITIES IN THE UNITED STATES

STATE	STATION[b]	HEATING DEGREE DAYS (yearly total)	DESIGN TEMPERATURES			DEGREES NORTH LATITUDE[c]
			Winter	Summer		
			97½%	Dry bulb 2½%	Wet bulb 2½%	
GA	Athens	2,929	22	92	77	34°00′
	Atlanta	2,961	22	92	76	33°40′
	Augusta	2,397	23	95	79	33°20′
	Columbus	2,383	24	93	78	32°30′
	Macon	2,136	25	93	78	32°40′
	Rome	3,326	22	93	78	34°20′
	Savannah	1,819	27	93	79	32°10′
HI	Hilo	0	62	83	74	19°40′
	Honolulu	0	63	86	75	21°20′
ID	Boise	5,809	10	94	66	43°30′
	Lewiston	5,542	6	93	66	46°20′
	Pocatello	7,033	-1	91	63	43°00′
IL	Chicago (Midway)	6,155	0	91	75	41°50′
	Chicago (O'Hare)	6,639	-4	89	76	42°00′
	Chicago[d]	5,882	2	91	77	41°50′
	Moline	6,408	-4	91	77	41°30′
	Peoria	6,025	-4	89	76	40°40′
	Rockford	6,830	-4	89	76	42°10′
	Springfield	5,429	2	92	77	39°50′
IN	Evansville	4,435	9	93	78	38°00′
	Fort Wayne	6,205	1	89	75	41°00′
	Indianapolis	5,699	2	90	76	39°40′
	South Bend	6,439	1	89	75	41°40′
IA	Burlington	6,114	-3	91	77	40°50′
	Des Moines	6,588	-5	91	77	41°30′
	Dubuque	7,376	-7	88	75	42°20′
	Sioux City	6,951	-7	92	77	42°20′
	Waterloo	7,320	-10	89	77	42°30′
KS	Dodge City	4,986	5	97	73	37°50′
	Goodland	6,141	0	96	70	39°20′
	Topeka	5,182	4	96	78	39°00′
	Wichita	4,620	7	98	76	37°40′
KY	Covington	5,265	6	90	75	39°00′
	Lexington	4,683	8	91	76	38°00′
	Louisville	4,660	10	93	77	38°10′
LA	Alexandria	1,921	27	94	79	31°20′
	Baton Rouge	1,560	29	93	80	30°30′
	Lake Charles	1,459	31	93	79	30°10′
	New Orleans	1,385	33	92	80	30°00′
	Shreveport	2,184	25	96	79	32°30′
ME	Caribou	9,767	-13	81	69	46°50′
	Portland	7,511	-1	84	72	43°40′
MD	Baltimore	4,654	13	91	77	39°10′
	Baltimore[d]	4,111	17	89	78	39°20′
	Frederick	5,087	12	91	77	39°20′

DEGREE DAY AND DESIGN TEMPERATURES[a] FOR CITIES IN THE UNITED STATES

STATE	STATION[b]	HEATING DEGREE DAYS (yearly total)	Winter 97$^1/_2$%	Summer Dry bulb 2$^1/_2$%	Summer Wet bulb 2$^1/_2$%	DEGREES NORTH LATITUDE[c]
MA	Boston	5,634	9	88	74	42°20′
	Pittsfield	7,578	-3	84	72	42°30′
	Worcester	6,969	4	84	72	42°20′
MI	Alpena	8,506	-6	85	72	45°00′
	Detroit (City)	6,232	6	88	74	42°20′
	Escanaba[d]	8,481	-7	83	71	45°40′
	Flint	7,377	1	87	74	43°00′
	Grand Rapids	6,894	5	88	74	42°50′
	Lansing	6,909	1	87	74	42°50′
	Marquette[d]	8,393	-8	81	70	46°30′
	Muskegon	6,696	6	84	73	43°10′
	Sault Ste. Marie	9,048	-8	81	70	46°30′
MN	Duluth	10,000	-16	82	70	46°50′
	Minneapolis	8,382	-12	89	5	44°50′
	Rochester	8,295	-12	87	75	44°00′
MS	Jackson	2,239	25	95	78	32°20′
	Meridian	2,289	23	95	79	32°20′
	Vicksburg[d]	2,041	26	95	80	32°20′
MO	Columbia	5,046	4	94	77	39°00′
	Kansas City	4,711	6	96	77	39°10′
	St. Joseph	5,484	2	93	79	39°50′
	St. Louis	4,900	6	94	77	38°50′
	St. Louis[d]	4,484	8	94	77	38°40′
	Springfield	4,900	9	93	77	37°10′
MT	Billings	7,049	-10	91	66	45°50′
	Great Falls	7,750	-15	88	62	47°30′
	Helena	8,129	-16	88	62	46°40′
	Missoula	8,125	-6	88	63	46°50′
NE	Grand Island	6,530	-3	94	74	41°00′
	Lincoln[d]	5,864	-2	95	77	40°50′
	Norfolk	6,979	-4	93	77	42°00′
	North Platte	6,684	-4	94	72	41°10′
	Omaha	6,612	-3	91	77	41°20′
	Scottsbluff	6,673	-3	92	68	41°50′
NV	Elko	7,433	-2	92	62	40°50′
	Ely	7,733	-4	87	59	39°10′
	Las Vegas	2,709	28	106	70	36°10′
	Reno	6,332	10	92	62	39°30′
	Winnemucca	6,761	3	94	62	40°50′
NH	Concord	7,383	-3	87	73	43°10′
NJ	Atlantic City	4,812	13	89	77	39°30′
	Newark	4,589	14	91	76	40°40′
	Trenton[d]	4,980	14	88	76	40°10′
NM	Albuquerque	4,348	16	94	65	35°00′
	Raton	6,228	1	89	64	36°50′
	Roswell	3,793	18	98	70	33°20′
	Silver City	3,705	10	94	64	32°40′

DEGREE DAY AND DESIGN TEMPERATURES[a] FOR CITIES IN THE UNITED STATES

STATE	STATION[b]	HEATING DEGREE DAYS (yearly total)	DESIGN TEMPERATURES			DEGREES NORTH LATITUDE[c]
			Winter	Summer		
			97½%	Dry bulb 2½%	Wet bulb 2½%	
NY	Albany	6,875	-1	88	74	42°50′
	Albany[d]	6,201	1	88	74	42°50′
	Binghamton	7,286	1	83	72	42°10′
	Buffalo	7,062	6	85	73	43°00′
	NY (Central Park)[d]	4,871	15	89	75	40°50′
	NY (Kennedy)	5,219	15	87	75	40°40′
	NY (LaGuardia)	4,811	15	89	75	40°50′
	Rochester	6,748	5	88	73	43°10′
	Schenectady[d]	6,650	1	87	74	42°50′
	Syracuse	6,756	2	87	73	43°10′
NC	Charlotte	3,181	22	93	76	35°10′
	Greensboro	3,805	18	91	76	36°10′
	Raleigh	3,393	20	92	77	35°50′
	Winston-Salem	3,595	20	91	75	36°10′
ND	Bismarck	8,851	-19	91	71	46°50′
	Devils Lake[d]	9,901	-21	88	71	48°10′
	Fargo	9,226	-18	89	74	46°50′
	Williston	9,243	-21	88	70	48°10′
OH	Akron-Canton	6,037	6	86	73	41°00′
	Cincinnati[d]	4,410	6	90	75	39°10′
	Cleveland	6,351	5	88	74	41°20′
	Columbus	5,660	5	90	75	40°00′
	Dayton	5,622	4	89	75	39°50′
	Mansfield	6,403	5	87	74	40°50′
	Sandusky[d]	5,796	6	91	74	41°30′
	Toledo	6,494	1	88	75	41°40′
	Youngstown	6,417	4	86	73	41°20′
OK	Oklahoma City	3,725	13	97	77	35°20′
	Tulsa	3,860	13	98	78	36°10′
OR	Eugene	4,726	22	89	67	44°10′
	Medford	5,008	23	94	68	42°20′
	Portland	4,635	23	85	67	45°40′
	Portland[d]	4,109	24	86	67	45°30′
	Salem	4,754	23	88	68	45°00′
PA	Allentown	5,810	9	88	75	40°40′
	Erie	6,451	9	85	74	42°10′
	Harrisburg	5,251	11	91	76	40°10′
	Philadelphia	5,144	14	90	76	39°50′
	Pittsburgh	5,987	5	86	73	40°30′
	Pittsburgh[d]	5,053	7	88	73	40°30′
	Reading[d]	4,945	13	89	75	40°20′
	Scranton	6,254	5	87	73	41°20′
	Williamsport	5,934	7	89	74	41°10′
RI	Providence	5,954	9	86	74	41°40′
SC	Charleston	2,033	27	91	80	32°50′
	Charleston[d]	1,794	28	92	80	32°50′
	Columbia	2,484	24	95	78	34°00′

DEGREE DAY AND DESIGN TEMPERATURES[a] FOR CITIES IN THE UNITED STATES

STATE	STATION[b]	HEATING DEGREE DAYS (yearly total)	Winter 97½%	Summer Dry bulb 2½%	Summer Wet bulb 2½%	DEGREES NORTH LATITUDE[c]
SD	Huron	8,223	-14	93	75	44°30'
	Rapid City	7,345	-7	92	69	44°00'
	Sioux Falls	7,839	-11	91	75	43°40'
TN	Bristol	4,143	14	89	75	36°30'
	Chattanooga	3,254	18	93	77	35°00'
	Knoxville	3,494	19	92	76	35°50'
	Memphis	3,232	18	95	79	35°00'
	Nashville	3,578	14	94	77	36°10'
TX	Abilene	2,624	20	99	74	32°30'
	Austin	1,711	28	98	77	30°20'
	Dallas	2,363	22	100	78	32°50'
	El Paso	2,700	24	98	68	31°50'
	Houston	1,396	32	94	79	29°40'
	Midland	2,591	21	98	72	32°00'
	San Angelo	2,255	22	99	74	31°20'
	San Antonio	1,546	30	97	76	29°30'
	Waco	2,030	26	99	78	31°40'
	Wichita Falls	2,832	18	101	76	34°00'
UT	Salt Lake City	6,052	8	95	65	40°50'
VT	Burlington	8,269	-7	85	72	44°30'
VA	Lynchburg	4,166	16	90	76	37°20'
	Norfolk	3,421	22	91	78	36°50'
	Richmond	3,865	17	92	78	37°30'
	Roanoke	4,150	16	91	74	37°20'
WA	Olympia	5,236	22	83	66	47°00'
	Seattle-Tacoma	5,145	26	80	64	47°30'
	Seattle[d]	4,424	27	82	67	47°40'
	Spokane	6,655	2	90	64	47°40'
WV	Charleston	4,476	11	90	75	38°20'
	Elkins	5,675	6	84	72	38°50'
	Huntington	4,446	10	91	77	38°20'
	Parkersburg[d]	4,754	11	90	76	39°20'
WI	Green Bay	8,029	-9	85	74	44°30'
	La Crosse	7,589	-9	88	75	43°50'
	Madison	7,863	-7	88	75	43°10'
	Milwaukee	7,635	-4	87	74	43°00'
WY	Casper	7,410	-5	90	61	42°50'
	Cheyenne	7,381	-1	86	62	41°10'
	Lander	7,870	-11	88	63	42°50'
	Sheridan	7,680	-8	91	65	44°50'

a. All data were extracted from the 1985 ASHRAE Handbook, Fundamentals Volume.
b. Design data developed from airport temperature observations unless noted.
c. Latitude is given to the nearest 10 minutes. For example, the latitude for Miami, Florida, is given as 25°50', or 25 degrees 50 minutes.
d. Design data developed from office locations within an urban area, not from airport temperature observations.

Chapter 17

EVALUATING SITES AND THEIR REQUIREMENTS

MULTIPLE-CHOICE EXAM

1. Which of the following will be looked at when evaluating a site?
    a. Slope   b. Properties
    c. Soil saturation   d. All of the above

2. Test data on soil based on undisturbed elements and establishing a vertical elevation reference point is commonly called:
    a. Permeability   b. Borings
    c. Benchmark   d. Soil conditions

3. After testing soil, a report on approved forms, must be filed with the appropriate agency within _____ days.
    a. 30   b. 25
    c. 20   d. 15

4. An area set aside for a replacement system is allowed to be used for which of the following?
    a. Parking lots   b. Buildings
    c. In-the-ground pools   d. None of these may be used

5. These may be used to make borings for soil tests.
    a. Backhoes   b. Power augers
    c. Hand augers   d. Both a and c

6. Any soil absorption site is required to have a minimum of _____ bore holes at the site.
    a. 4   b. 3
    c. 2   d. 5

7. Which of the following is a type of horizon for soil appearance?
    a. Color   b. Size
    c. Shape   d. Ledge

8. Land slopes that are more than _____ percent is not usually acceptable for a conventional disposal system.
    a. 15   b. 25
    c. 20   d. 30

9.   Perk tests are usually conducted by which of the following:

    a.   The homeowner   b.   A licensed professional

    c.   The code officer   d.   The D.E. P

10.   For what purpose is the sides of test holes scratched or roughed up?

    a.   To check for rocks   b.   To remove loose dirt

    c.   To see the color   d.   To expose natural soil

TRUE-FALSE EXAM

1.   The monitoring of a system may be required by the code officer.

    True   False

2.   Flood hazard sites are the best conditions in which to install soil absorption systems.

    True   False

3.   Any loose soil or material must be left in the test hole for greater absorption.

    True   False

4.   Soil mottles should be checked for when ground water is found.

    True   False

5.   Soil testing is easy; anyone can perform one.

    True   False

6.   Borings must be accurately referenced to the horizontal elevation and vertical reference points.

    True   False

7.   A code officer could require a detailed soil map.

    True   False

8.   Private sewers should always be installed below driveways and parking lots.

    True   False

9.   Even with evidence, you may never take a property owner at his word in regards to a suitable existing site.

    True   False

10. Sites with soils finer than sand or loamy sand can usually be approved for a private sewage disposal system.

True   False

11. Soil absorption systems must not be installed in a filled area, unless written approval is received.

True   False

12. The placement of fill is to be inspected by a local code officer.

True   False

13. The minimum horizontal distance for a seepage pit and a spring is 50 feet.

True   False

14. The bottom of a test hole is to be covered with two inches of fine sand.

True   False

15. Subsurface soil absorption systems are not allowed to be installed in alluvial or colluvial deposits that have shallow depths.

True   False

MULTIPLE-CHOICE ANSWERS

1.	d	3.	a	5.	d	7.	a	9.	b
2.	c	4.	d	6.	b	8.	c	10.	d

TRUE-FALSE ANSWERS

1.	True	5.	False	9.	False	13.	True
2.	False	6.	False	10.	False	14.	False
3.	False	7.	True	11.	True	15.	False
4.	True	8.	False	12.	True		

TABLE 17.1   Minimum horizontal separation distances for soil absorption systems. *Copyright 2006, International Code Council, Inc., Falls Church, Virginia. Reproduced with permission. All rights reserved.*

ELEMENT	DISTANCE (feet)
Cistern	50
Habitable building, below-grade foundation	25
Habitable building, slab-on-grade	15
Lake, high-water mark	50
Lot line	5
Reservoir	50
Roadway ditches	10
Spring	100
Streams or watercourse	50
Swimming pool	15
Uninhabited building	10
Water main	50
Water service	10
Water well	50

For SI: 1 foot = 304.8 mm.

Chapter 18
MATERIALS

Choosing the proper plumbing materials for various types of jobs can be complicated. With the wide variety of materials available for different types of plumbing work, keeping all the code requirements clear can be troublesome. Not only do you have to know what types of materials are acceptable for certain types of plumbing applications, but you should also be able to determine which of the approved materials is the best choice for the given job. For example, both galvanized-steel pipe and Schedule-40 plastic pipe are approved materials for most types of drainage. Given the job of choosing which of these two materials to use for a common residential sink drain, which would you choose?

Either galvanized steel or Schedule-40 plastic pipe will earn you a correct answer on your plumbing exam. But in the field, Schedule-40 plastic pipe would be the best choice. Galvanized-steel pipe tends to rust. The rust creates rough spots on interior sections of the pipe. In time, these rough sections will act as a trap for hair, grease, and other unwanted debris. With a little more time, the pipe will begin to close up, until eventually the interior of the pipe is obstructed. When this happens, water cannot drain past the clog, and the wastewater backs up into the sink.

The threaded joints of galvanized pipe are another trouble spot. Since the wall thickness of the pipe is minimal where the threads have been cut, leaks often occur around the threaded sections. This is a result of corrosion.

While plastic pipes occasionally become clogged, they do so much less frequently than galvanized-steel pipes. It is possible for a joint connection between plastic pipe to go bad and pull apart, but there are no threads to create weak spots. Since plastic is resistant to most corrosive action and rust, it does not deteriorate like steel pipe can. These factors combine to make plastic pipe a better all-round choice for drainage pipe.

MULTIPLE-CHOICE EXAM

1. Which of the following, when used in a plumbing system, must have a cast, embossed, stamped, or indelibly marked symbol of the manufacturer:

    a. Pipe
    b. Pipe fittings
    c. Traps
    d. Fixtures
    e. All of the above
    f. None of the above

2. Water-service pipe must be resistant to which of the following:

    a. Corrosive action
    b. Degrading action
    c. Both A and B
    d. None of the above

3. Water-distribution pipe must be resistant to which of the following:

    a. Corrosive action
    b. Degrading action
    c. Both A and B
    d. None of the above

4. When detrimental conditions are suspected to exist in an area where a water service is to be installed, what must be done:

    a. Copper pipe must be used
    b. Polybutylene pipe must be used
    c. A chemical analysis of the soil must be made
    d. Any of the above

5. When PE plastic piping is used as a water service, the plastic piping may not extend farther than _____ within a building:

    a. 5 feet
    b. 8 feet
    c. 10 feet
    d. None of the above

6. Which of the following materials may be used as a water service:

    a. Copper pipe
    b. PEX pipe
    c. Lead pipe
    d. a and b
    e. All of the above

7. Galvanized-steel pipe may be used for which of the following purposes:

    a. Water service
    b. Water distribution
    c. Above-ground vents
    d. A and B
    e. A, B, and C

8. Pipe or tubing used to convey hot water must have a minimum pressure rating of _____ when rated at 180 degrees F:

   a. 60 psi   b. 80 psi

   c. 100 psi   d. 125 psi

9. A building sewer pipe may be made of which of the following materials?

   a. Vitrified clay pipe   b. Polybutylene pipe

   c. Polyethylene pipe   d. Any of the above

10. If DWV copper is used as a building sewer material, what "Type" of copper must it be?

    a. Type M   b. Type L

    c. Type K   d. Any of the above

    e. A and C   f. B and C

11. If DWV copper is used as an underground building drainage or vent material, what "Type" of copper must it be?

    a. Type M   b. Type L

    c. Type K   d. DWV

    e. Any of the above   f. A and C

    g. B and C

12. Can a building sewer pipe be installed in a trench where a water service is installed?

    a. No, never

    b. Yes, always

    c. Yes, under certain circumstances

    d. Yes, but only if the sewer is made of cast iron

13. Materials used to vent or drain chemical waste must be:

    a. Made of copper

    b. Made of cast iron

    c. Resistant to corrosion and degradation for the chemicals involved

    d. Made of stainless steel

14. Pipe that is open-jointed, horizontally split, or perforated may be used for which of the following purposes:

    a. Sanitary vents   b. Sanitary drains

    c. Subsoil drains   d. Any of the above

    e. None of the above

15. Threaded drainage-pipe fittings shall be of which of the following types:

   a. Solvent-weld type
   b. Exposed-drainage type
   c. Recessed-drainage type
   d. None of the above

16. Fittings must be made in such a way that the flow of water through them is not retarded or obstructed by which of the following:

   a. Ledges
   b. Shoulders
   c. Reductions
   d. All of the above
   e. None of the above

17. Pipe fittings may be made of which of the following materials:

   a. Cast iron
   b. Copper
   c. Plastic
   d. All of the above
   e. A and B only

18. Which of the following materials may cleanout plugs be made of:

   a. Lead
   b. Brass
   c. Copper
   d. All of the above

19. Which of the following materials may cleanout plugs be made of:

   a. Lead
   b. Copper
   c. Plastic
   d. All of the above

20. When use of a raised-head type of cleanout plug may create a tripping hazard, what type of head may be used in its place:

   a. A countersunk head
   b. A flush head
   c. Either A or B
   d. Neither A or B

21. What type of surface must plumbing fixtures have:

   a. White surfaces
   b. Stainless-steel surfaces
   c. Smooth, impervious surfaces
   d. Porous surfaces

22. What type of surface must plumbing fixtures not have:

   a. Stainless-steel surfaces
   b. Smooth, impervious surfaces
   c. Concealed fouling surfaces
   d. China surfaces

23.   A water-service pipe may not be installed in ground that is:

     a.   Contaminated with solvents   b.   Contaminated with fuels

     c.   Corrosive   d.   All of the above

24.   At what temperature is the pressure rating of a water-service pipe measured:

     a.   72 degrees F.   b.   73.4 degrees F.

     c.   76.3 degrees F.   d.   80 degrees F.

25.   Where water pressure exceeds 160 psi, piping material for a water service must be:

     a.   Copper

     b.   Polybutylene

     c.   Rated with a minimum working pressure equal to the highest available pressure

     d.   Any of the above

26.   If a construction joint is used, it must have a keyway in this section of the joint.

     a.   Upper   b.   Middle

     c.   Lower   d.   Inside

27.   Steel tanks must conform to this:

     a.   UL 65   b.   UV 70

     c.   UV 65   d.   UL 70

28.   For a 1,001 to 1,250 gallon tank of vertical cylindrical design, the minimum gage thickness for the complete tank is:

     a.   10 gage   b.   12 gage

     c.   15 gage   d.   None of the above

29.   For a 500 to 1,000 gallon tank of vertical cylindrical design, the minimum diameter of the bottom and sidewall is:

     a.   54 inch   b.   65 inch

     c.   None   d.   50 inch

30.   Joints made of _____ must be tested and approved before they can be painted or varnished.

     a.   lead   b.   copper

     c.   pvc   d.   cast-iron

31. These types of joints must be made on surfaces that are clean and dry?

    a. Heat-fusion
    b. Cement
    c. Concrete
    d. Mechanical

32. A continuous water stop or baffle must have a minimum width of _____ inches.

    a. 44
    b. 56
    c. 40
    d. 46

33. A baffle must be constructed from which of the following materials:

    a. Copper
    b. a and d
    c. Iron
    d. Rubber

34. These types of joints can be made below or above ground

    a. Mechanical joints
    b. Flared joints
    c. Solvent cement joints
    d. Compression joints

35. A mechanical joint cannot be:

    a. Caulked
    b. Threaded
    c. Welded
    d. All of these

   Congratulations, you have just completed the multiple-choice exam on approved materials. Before you turn to the back of the chapter to check your answers, take the true-false test that follows this paragraph.

TRUE-FALSE EXAM

1. If copper tubing is used for underground drainage piping, it must have a weight not less than that of DWV copper.

    True   False

2. Copper tubing used for above-ground drainage must have a weight of not less than that of Type-M copper.

    True   False

3. Copper tubing used for water piping must have a weight of not less than that of Type K copper.

    True   False

4.   All hard-drawn copper tubing must be marked by means of a continuous and indelibly colored stripe at least 1/4 inch wide.

True   False

5.   Type L copper is marked with a blue line.

True   False

6.   Cleanouts must be designed to be gas- and watertight without the use of any gasket, packing, or washer.

True   False

7.   Type M copper is marked with a red line.

True   False

8.   Gate valves used on drainage piping must be of a full-way design and of a corrosive-resistant material.

True   False

9.   Fittings used for drainage must be designed to allow a drainage pitch of at least 1/8 inch per each foot of fall.

True   False

10.   A pressure-reducing valve must be designed to remain open to permit uninterrupted water flow in case of value failure.

True   False

11.   Code requires the manufacturer's mark or name and the quality of the product or identification to be embossed on each length of pipe.

True   False

12.   All tanks used, must be marked with their capacity.

True   False

13.   Solvent cement joints must be made while the joint is damp.

True   False

14.   Concrete pipe joints are to be made with elastomeric seals.

True   False

15.   Compression joints that are approved can be used to join cast-iron pipes.

True   False

16.   A continuous water stop must be set horizontally in the joint.

True   False

17. The joint between a concrete tank must be sealed water tight with the use of super glue.

    True   False

18. A 500 to 1,000 gallon tank that is of horizontal cylindrical must have a minimum diameter of 54 inches.

    True   False

19. Pipes for private sewage disposal systems are required to have rough walls.

    True   False

20. Mechanical joints are not allowed on underground plumbing unless approved.

    True   False

21. The collars for manholes and their extensions must be made with the same material as the tanks that they are being used with.

    True   False

22. Mastic or hot-pour bituminous joints are prohibited.

    True   False

23. Joints between a concrete tank and tank cover must be of tongue-and-cheek type.

    True   False

24. Heat-fusion joints must be left undisturbed until it cools.

    True   False

25. It is common to use solvent-cement joints between different types of plastic pipe.

    True   False

MULTIPLE-CHOICE ANSWERS

1.	e	8.	c	15.	c	22.	c	29.	c
2.	c	9.	a	16.	d	23.	d	30.	d
3.	c	10.	f	17.	d	24.	b	31.	a
4.	c	11.	e	18.	d	25.	c	32.	b
5.	a	12.	c	19.	d	26.	c	33.	b
6.	d	13.	c	20.	a	27.	d	34.	c
7.	e	14.	c	21.	c	28.	a	35.	d

TRUE-FALSE ANSWERS

1.	False	8.	True	14.	True	20.	False
2.	False	9.	True	15.	True	21.	True
3.	False	10.	True	16.	False	22.	True
4.	True	11.	True	17.	False	23.	False
5.	True	12.	True	18.	True	24.	True
6.	True	13.	False	19.	False	25.	False
7.	True						

TABLE 18.1 Tank capacity. *Copyright 2006, International Code Council, Inc., Falls Church, Virginia. Reproduced with permission. All rights reserved.*

TANK DESIGN AND CAPACITY		MINIMUM GAGE THICKNESS	MINIMUM DIAMETER
Vertical cylindrical			
500 to 1,000 gallons	Bottom and sidewalls	12 gage	None
	Cover	12 gage	
	Baffles	12 gage	
1,001 to 1,250 gallons	Complete tank	10 gage	None
1,251 to 1,500 gallons	Complete tank	7 gage	None
Horizontal cylindrical			
500 to 1,000 gallons	Complete tank	12 gage	54-inch diameter
1,001 to 1,500 gallons	Complete tank	12 gage	64-inch diameter
1,501 to 2,500 gallons	Complete tank	10 gage	76-inch diameter
2,501 to 9,000 gallons	Complete tank	7 gage	76-inch diameter
9,001 to 12,000 gallons	Complete tank	$1/4$-inch plate	None
Over 12,000 gallons	Complete tank	$5/16$ inch	None

For SI: 1 inch = 25.4 mm, 1 gallon = 3.785 L.

TABLE 18.2   Private sewage disposal system pipe. *Copyright 2006, International Code Council, Inc., Falls Church, Virginia. Reproduced with permission. All rights reserved.*

MATERIAL	STANDARD
Acrylonitrile butadiene styrene (ABS) plastic pipe	ASTM D 2661; ASTM D 2751; ASTM F 628
Asbestos-cement pipe	ASTM C 428
Cast-iron pipe	ASTM A 74; ASTM A 888; CISPI 301
Coextruded composite ABS DWV Schedule 40 IPS pipe (solid)	ASTM F 1488; ASTM F 1499
Coextruded composite ABS DWV Schedule 40 IPS pipe (cellular core)	ASTM F 1488; ASTM F 1499
Coextruded composite ABS sewer and drain DR-PS in PS35, PS50, PS100, PS140 and PS200	ASTM F 1488; ASTM F 1499
Coextruded composite PVC DWV Schedule 40 IPS pipe (solid)	ASTM F 1488
Coextruded composite PVC DWV Schedule 40 IPS pipe (cellular core)	ASTM F 1488
Coextruded composite PVC-IPS-DR of PS140, PS200, DWV	ASTM F 1488
Coextruded composite PVC 3.25 OD DWV pipe	ASTM F 1488
Coextruded composite PVC sewer and drain DR-PS in PS35, PS50, PS100, PS140 and PS200	ASTM F 1488
Concrete pipe	ASTM C 14; ASTM C 76; CSA A257.1M; CSA A257.2M
Copper or copper-alloy tubing (Type K or L)	ASTM B 75; ASTM B 88; ASTM B 251
Polyvinyl chloride (PVC) plastic pipe (Type DWV, SDR26, SDR35, SDR41, PS50 or PS100)	ASTM D 2665; ASTM D 2949; ASTM D 3034; ASTM F 891; CSA B182.2; CSA B182.4
Vitrified clay pipe	ASTM C 4; ASTM C 700

Chapter 19

SIZING AND INSTALLING SOIL ABSORPTION SYSTEMS

FILL IN THE BLANK EXAM

1.   The sizing of a soil absorption system is usually done by _____.

2.   The minimum spacing between trenches is _____ feet.

3.   A maximum length for a trench is _____ feet.

4.   _____ must not begin if the soil is so wet that it can be rolled between your hands.

5.   Soiled absorption systems must not be installed in _____ ground.

6.   Seepage trench excavations are required to be one foot to _____ feet wide.

7.   The bottom of a trench or bed excavation must be _____.

8.   The _____ must be installed in 2-inch layers over the top of all distribution piping.

9.   Any distribution, _____, must be solid-wall pipe.

10.   The _____ of the distribution pipe must not be less than eight inches below the original surface in continuous straight or curved lines.

Cross off the answers as you use them

frozen	an engineer	PVC	top	Excavation
aggregrate	100	six	five	level

TRUE-FALSE EXAM

1. The absorption area of a seepage trench must be computed only by the bottom of the trench area.
    True   False

2. Seepage pits must have a minimum inside diameter of seven feet.
    True   False

3. Observation pipes must be located 25 feet from any window, door, or air intake of any building that is used for human occupancy.
    True   False

4. The first foot of backfill material must be solid.
    True   False

5. Frozen soil is the best type of material to be used for backfill.
    True   False

6. Up to four distribution pipes can be served by a single, 4-inch, observation, with one exception.
    True   False

7. Building paper is considered to be an ideal cover material.
    True   False

8. The minimum amount of soil needed for covering is 18 inches.
    True   False

9. The bottom of a pit must be left open to the soil below.
    True   False

10. Tank capacity and design specifications can be established by checking the table in your local code book.
    True   False

11. Piping installed in a seepage bed must be evenly spaced a maximum of five feet and a minimum of three feet apart.
    True   False

12. All distribution piping must have a slope of at least one foot for every 25 feet of pipe that is installed.
    True   False

13. The words slope and grade are interchangeable and are similar in meaning.

     True   False

14. The bottom 12 inches of an observation pipe must be perforated and extend to the bottom of he aggregate.

     True   False

15. Aggregate must be covered with an approved synthetic material or nine inches of straw.

     True   False

FILL IN THE BLANK ANSWERS

1.	an engineer	6.	five
2.	six	7.	level
3.	100	8.	aggregate
4.	Excavation	9.	PVC
5.	frozen	10.	Top

TRUE-FALSE ANSWERS

1.	True	5.	False	9.	True	13.	True
2.	False	6.	True	10.	True	14.	True
3.	True	7.	False	11.	True	15.	True
4.	False	8.	True	12.	False		

TABLE 19.1 Minimum absorption area for one- and two-family dwellings. *Copyright 2006, International Code Council, Inc., Falls Church, Virginia. Reproduced with permission. All rights reserved.*

PERCOLATION CLASS	PERCOLATION RATE (minutes required for water to fall 1 inch)	SEEPAGE TRENCHES OR PITS (square feet per bedroom)	SEEPAGE BEDS (square feet per bedroom)
1	0 to less than 10	165	205
2	10 to less than 30	250	315
3	30 to less than 45	300	375
4	45 to 60	330	415

For SI: 1 minute per inch = 2.4 s/mm, 1 square foot = 0.0929 m^2.

TABLE 19.2 Effective square-foot absorption area for seepage pits. *Copyright 2006, International Code Council, Inc., Falls Church, Virginia. Reproduced with permission. All rights reserved.*

INSIDE DIAMETER OF CHAMBER IN FEET PLUS 1 FOOT FOR WALL THICKNESS PLUS 1 FOOT FOR ANNULAR SPACE	DEPTH IN FEET OF PERMEABLE STRATA BELOW INLET					
	3	4	5	6	7	8
7	47	88	110	132	154	176
8	75	101	126	151	176	201
9	85	113	142	170	198	226
10	94	126	157	188	220	251
11	104	138	173	208	242	277
13	123	163	204	245	286	327

For SI: 1 foot = 304.8 mm.

—NOTES—

Chapter 20

PRESSURE DISTRIBUTION SYSTEMS

Plumber's Licensing Study Guide

FILL IN THE BLANK EXAM

1.   If a system receives more than _____ gallons of effluent, there is likely to be a need for two systems.

2.   The estimated wastewater from a typical residential dwelling is estimated at _____ gallons per bedroom.

3.   For a distribution pipe that is 4 inches in diameter, you need _____ inches of suitable soil.

4.   Sizing these systems is best left up to whom?

5.   _____ is the biggest obstacle in sizing a system yourself.

6.   If the percolation rate is zero to less than 10, what is the design loading factor?

7.   The design load factor is 0.4; what would this make the percolation rate?

8.   What is the minimum capacity for a single system?

9.   Besides math, which else is required to size a system?

10. Pressure distribution systems are _____ on any site that meets the requirements for a conventional soil absorption system.

Cross off the answers as you use them

Math	5,000	design	professionals	1.2	75%
150	53	formulas	permissible		

FILL IN THE BLANK ANSWERS

1. 5,000
2. 150
3. 53
4. design professionals
5. math

6. 1.2
7. 45–60
8. 75 %
9. formulas
10. permissible

TABLE 20.1   Soil required. *Copyright 2006, International Code Council, Inc., Falls Church, Virginia. Reproduced with permission. All rights reserved.*

DISTRIBUTION PIPE (inches)	SUITABLE SOIL (inches)
1	49
2	50
3	52
4	53

For SI: 1 inch = 25.4 mm.

TABLE 20.2   Design loading rate. *Copyright 2006, International Code Council, Inc., Falls Church, Virginia. Reproduced with permission. All rights reserved.*

PERCOLATION RATE (minutes per inch)	DESIGN LOADING FACTOR (gallons per square foot per day)
0 to less than 10	1.2
10 to less than 30	0.8
30 to less than 45	0.72
45 to 60	0.4

For SI: 1 minute per inch = 2.4 s/mm, 1 gallon per square foot = 0.025 L/m^2.

Chapter 21
TANKS

FILL IN THE BLANK EXAM

1.   Septic tanks are required to have two _____.

2.   A _____ bedroom home requires a tank with the capacity of 1,000 gallons.

3.   Tanks installed in ground water must be _____.

4.   All tanks must be _____.

5.   A septic tank must be made from an approved material, such as _____ steel.

6.   The bottom of an outlet opening shall be a minimum of _____ inches lower than the bottom of an inlet.

7.   One _____ opening is required for each tank.

8.   _____ restoration for private sewage disposal systems is not allowed unless otherwise approved.

9.   A pumping and _____ schedule for each holding tank installation shall be submitted to the local code officer.

10. A _____ tank must not be installed closer than 20 feet from any part of a building.

Cross off the answers as you use them

Chemical	compartments	holding	inspection	watertight
Anchored	welded	two	maintenance	three

TRUE-FALSE EXAM

1. If the system receives more than 5,000 gallons of effluent, there is likely to be a need for two systems?
    True   False

2. The liquid depth of a tank shall not be less than 30 inches and a maximum average of six feet.
    True   False

3. A minimum of one inch of clear space must be provided over the top of the baffles or tees.
    True   False

4. One inspection opening is required for each tank.
    True   False

5. When inspection pipes terminate above finished grade, the height of the pipe must not be more than four inches.
    True   False

6. The requirements for a treatment tank differ from those needed for a septic tank.
    True   False

7. It is encouraged to install more than four tanks in a series.
    True   False

8. There is a table in the code book that lists examples for sizing purposes.
    True   False

9. If a septic tank is used as a holding tank, the outlets do not need to be sealed.
    True   False

10. All service ports must be equipped with either a locking cover or a brass cleanout plug.
    True   False

11. The secondary compartment of a septic tank shall not have less than a capacity of 250 gallons.
    True   False

12. Fiberglass tanks require a 2-inch fiberglass collar.
    True   False

13. A minimum of two inches of clear space must be provided over the top of the baffles or tees.
    True   False

14. Every compartment of a tank requires a manhole.

True   False

15. Holding tanks must be sized to have a minimum of a 7-day holing capacity.

True   False

FILL IN THE BLANK ANSWERS

1. compartments	6. two
2. three	7. inspection
3. anchored	8. Chemical
4. watertight	9. maintenance
5. welded	10. holding

TRUE-FALSE ANSWERS

1. True	5. False	9. False	13. True
2. True	6. False	10. True	14. True
3. False	7. False	11. False	15. False
4. True	8. True	12. True	

TABLE 21.1   Septic tank capacity for one- and two-family dwellings. *Copyright 2006, International Code Council, Inc., Falls Church, Virginia. Reproduced with permission. All rights reserved.*

NUMBER OF BEDROOMS	SEPTIC TANK (gallons)
1	750
2	750
3	1,000
4	1,200
5	1,425
6	1,650
7	1,875
8	2,100

For SI: 1 gallon = 3.785 L.

TABLE 21.2   Minimum horizontal separation distances for treatment tanks. *Copyright 2006, International Code Council, Inc., Falls Church, Virginia. Reproduced with permission. All rights reserved.*

ELEMENT	DISTANCE (feet)
Building	5
Cistern	25
Foundation wall	5
Lake, high water mark	25
Lot line	2
Pond	25
Reservoir	25
Spring	50
Stream or watercourse	25
Swimming pool	15
Water service	5
Well	25

For SI: 1 foot = 304.8 mm.

Chapter 22

MOUND SYSTEMS

TRUE-FALSE EXAM

1.   With the right engineer, a mound system can make almost any piece of land a suitable septic site.
      True   False

2.   Mound systems are a way of dealing with difficult soils.
      True   False

3.   The minimum soil depth with a restricting factor of Pervious rock is 60 inches.
      True   False

4.   It is not wise to use a mound system when you have a slow perk rate.
      True   False

5.   You definitely want an experienced professional to design a mound system.
      True   False

6.   With a restricting factor such as rock fragments, the minimum soil depth is 12 inches.
      True   False

7.   A one-bedroom home on a 4 percent slope with a loading rate of 150 gallons per day requires the trench width to be 42 inches.
      True   False

8.   Bedrock is an ideal spot for a septic system.
      True   False

9.   A mound system is most often used when a conventional system will not work due to soil conditions.
      True   False

10.   When working with a standard table like those in most codebooks, they are just used as guidelines.
      True   False

TRUE-FALSE ANSWERS

1. True
2. True
3. False

4. False
5. True
6. False

7. False
8. False

9. True
10. False

TABLE 22.1 Minimum soil depths for mound system installation. *Copyright 2006, International Code Council, Inc., Falls Church, Virginia. Reproduced with permission. All rights reserved.*

RESTRICTING FACTOR	MINIMUM SOIL DEPTH TO RESTRICTION (inches)
High ground water	24
Impermeable rock strata	60
Pervious rock	24
Rock fragments (50-percent volume)	24

For SI: 1 inch = 25.4 mm.

TABLE 22.2 Design criteria for a mound for a one-bedroom home on a 0- to 6-percent slope with loading rates of 150 gallons per day for slowly permeable soil. *Copyright 2006, International Code Council, Inc., Falls Church, Virginia. Reproduced with permission. All rights reserved.*

	DESIGN PARAMETER	SLOPE (percent)			
		0	2	4	6
A	Trench width, feet	3	3	3	3
B	Trench length, feet	42	42	42	42
	Number of trenches	1	1	1	1
D	Mound height, inches	12	12	12	12
F	Mound height, inches	9	9	9	9
G	Mound height, inches	12	12	12	12
H	Mound height, inches	18	18	18	18
I	Mound width, feet[a]	15	15	15	15
J	Mound width, feet[a]	11	8	8	8
K	Mound length, feet	10	10	10	10
L	Mound length, feet	62	62	62	62
P	Distribution pipe length, feet	20	20	20	20
	Distribution pipe diameter, inches	1	1	1	1
	Number of holes per distribution pipes[b]	9	9	9	9
	Hole spacing, inches[b]	30	30	30	30
	Hole diameter, inches[b]	0.25	0.25	0.25	0.25
W	Mound width, feet	25	26	26	26

For SI: 1 inch = 25.4 mm, 1 foot = 304.8 mm, 1 gallon = 3.785 L.

a. Additional width to obtain required basal area.

b. Last hole is located at the end of the distribution pipe, which is 15 inches from the other hole.

TABLE 22.3 Design criteria for a two-bedroom home for a mound on a 0- to 6-percent slope with loading rates of 300 gallons per day for slowly permeable soil. *Copyright 2006, International Code Council, Inc., Falls Church, Virginia. Reproduced with permission. All rights reserved.*

	DESIGN PARAMETER	SLOPE (percent)			
		0	2	4	6
A	Trench width, feet	3	3	3	3
B	Trench length, feet	42	42	42	42
	Number of trenches	2	2	2	2
C	Trench spacing, feet	15	15	15	15
D	Mound height, inches	12	12	12	12
E	Mound height, inches	12	17	25	25
F	Mound height, inches	9	9	9	9
G	Mound height, inches	12	12	12	12
H	Mound height, inches	18	18	18	18
I	Mound width, feet[a]	12	20	20	20
J	Mound width, feet	12	8	8	8
K	Mound length, feet	10	10	10	10
L	Mound length, feet	62	62	62	62
P	Distribution pipe length, feet	20	20	20	20
	Distribution pipe diameter, inches	1	1	1	1
	Number of holes per distribution pipe[b]	9	9	9	9
	Hole spacing, inches[b]	30	30	30	30
	Hole diameter, inches	0.25	0.25	0.25	0.25
R	Manifold length, feet	15	15	15	15
	Manifold diameter, inches[c]	2	2	2	2
W	Mound width, feet	42	46	46	46

For SI: 1 inch = 25.4 mm, 1 foot = 304.8 mm, 1 gallon = 3.785 L.

a. Additional width to obtain required basal area.

b. Last hole is located at the end of the distribution pipe, which is 15 inches from the other hole.

c. Diameter dependent on the size of pipe from pump and inlet position.

TABLE 22.4 Design criteria for a three-bedroom home for a mound on a 0- to 6-percent slope with loading rates of 450 gallons per day for slowly permeable soil. *Copyright 2006, International Code Council, Inc., Falls Church, Virginia. Reproduced with permission. All rights reserved.*

	DESIGN PARAMETER	SLOPE (percent)			
		0	2	4	6
A	Trench width, feet	3	3	3	3
B	Trench length, feet	63	63	63	63
	Number of trenches	2	2	2	2
C	Trench spacing, feet	15	15	15	15
D	Mound height, inches	12	12	12	12
E	Mound height, inches	12	17	20	25
F	Mound height, inches	9	9	9	9
G	Mound height, inches	12	12	12	12
H	Mound height, inches	18	18	18	18
I	Mound width, feet[a]	12	20	20	20
J	Mound width, feet[a]	12	8	8	8
K	Mound length, feet	10	10	10	10
L	Mound length, feet	62	62	62	62
P	Distribution pipe length, feet	31	31	31	31
	Distribution pipe diameter, inches	1¼	1¼	1¼	1¼
	Number of holes per distribution pipe[b]	13	13	13	13
	Hole spacing[b], inches	30	30	30	30
	Hole diameter, inches	0.25	0.25	0.25	0.25
R	Manifold length, feet	15	15	15	15
	Manifold diameter, inches[c]	2	2	2	2
W	Mound width, feet	42	46	46	46

For SI: 1 inch = 25.4 mm, 1 foot = 304.8 mm, 1 gallon = 3.785 L.
a. Additional width to obtain required basal area.
b. First hole is located 12 inches from the manifold.
c. Diameter dependent on the size of pipe from pump and inlet position.

STUDY TIPS

There are many options available to people who wish to become licensed plumbers. If you want to earn a plumber's license, you will have to learn the plumbing code. This can be done by attending several years of vocational school, or by working in the field and studying on your own. Both ways are effective, and different people will share different opinions on which method is best suited to their personality.

When I started in the plumbing trade, I was a plumber's helper. A laborer would have been a better choice for my title, based on the type of work I did. My first six months of plumbing consisted of digging ditches, operating jackhammers, and carrying pipe, cast-iron pipe at that. During those six months, I lost 60 pounds, came home so tired I could hardly walk up the stairs, and nearly quit three times. It was difficult to be but so impressed with the plumbing trade when it seemed all I did was grunt work.

When it was time for my six-month review, my supervisor and the company manager met with me. They presented me with my performance report and I was impressed. It was then that I found out that the company tried to make life as hard as possible for rookies during their first six months of employment. The company wanted to make sure that the people they chose to train as plumbers had the grit and determination to stick with the program. I passed muster. From that day on, my plumbing career took turns for the better.

In the second half of my first year as a plumber's helper, I learned to solder joints, cut and thread pipe, pour lead joints, and in general, how to install complete plumbing systems. My job sites ranged from remote residential jobs to a large health spa. The experience I gained gave me a sense of pride. By the end of my first year in the trade, I could plumb a new house without any help.

After I was taken out of the ditches and off jackhammer duty, I realized that I really enjoyed plumbing. The strange thing is, I still do, and it's been 25 years since I picked up my first plumbing shovel. Anyway, once I knew I wanted to be a plumber, I bought a code book. It was expensive, but I wanted to learn how to be the best plumber in town, and I would have to know and understand the code in order to achieve my goal.

My code book went with me almost everywhere I went. When I would get to work early, I would sit in my truck and study the code. When I was riding to job sites with the plumber who was training me, his name was Jerry, I would ask him endless questions about the code. You see, the language in the code book was confusing to me at that time, and it helped a lot to have a knowledgeable plumber who would explain it to me.

I read my code book during lunch breaks and on the way back to the shop, when I could get Jerry to drive. When I would get home, I'd take a hot bath and read more of the code. You could probably build a case to show that I was obsessed with the plumbing code. I knew it was my ticket to making more money and having better working conditions. Of course, 25 years ago, I had no idea how good the plumbing trade was going to be to me. But, I can tell you know, it has been very good to me.

What does a plumber who has put in 25 years of field service do? Well, I write books, build houses now and then, and dabble in real estate, I offer training and development courses, and, of course, I still run a plumbing business. Do I still do plumbing myself? Yes, I do. Do I have to go out and do plumbing with my own hands? No, I could make enough money to be comfortable without touching a piece of pipe, but I enjoy plumbing, so I still get out there and do it.

I don't know your reasons for wanting to become a master plumber, but if you're reading this book, I assume you are interested in obtaining a master's license. Not everyone who comes into the trade enjoys the work as much as I do, but most of the plumbers I've known have been satisfied with their career choices. One of the questions I get asked most often is that of how much money does a plumber really make. This is a difficult question to answer, since not all plumbers make the same amounts of money. I can say this, if you are a good, experienced plumber, you can make a better-than-average living, and there is almost always someplace where you can work, even in the worst economic times.

SELF-STUDY

Self-study is the least expensive way to prepare for your licensing exam. Code-preparation classes, while not extremely expensive, cost money. With a local code book and this book, you can teach yourself the plumbing code. Your code book will provide the facts and foundation of the plumbing code, and this book will test your knowledge to expose areas which you are weak in.

CODE CLASSES

Code classes exist to help plumbers and aspiring plumbers to learn and understand the code. If the instructor of the class is good, the price of admission is money well spent. There is no substitute for having an experienced instructor available to question about confusion which may arise from the plumbing code.

When I was learning the trade, code classes didn't exist, at least not in my area or to my knowledge. If code-preparation classes had been available to me, I would have attended. I don't know what the national average is for people passing or failing plumbing licensing exams. Experience has shown me that a lot of plumbers and would-be plumbers fail the exams on their first attempts. I passed all of my tests on the first try, and the students who have attended my code classes have done very well in passing their tests on the first run.

Code classes are not a prerequisite for preparing to take a licensing exam, but I feel they are a good value and an investment in your plumbing future. If you combine a good code class with extensive self-study, you can hardly lose.

THE BUDDY SYSTEM

If you can't afford a code class, perhaps you can study with the buddy system. If you know someone else who is preparing for a licensing exam, the two of you can work together. Take turns being instructor and student. Test each others knowledge with the exams in this book. Working together can achieve your goal quicker than working alone, and the competitiveness between you and your partner may make you put a little extra effort into your studies.

LEARN AND UNDERSTAND IT

The key to passing a licensing exam is the plumbing code; you must learn it and understand it. Learning the code is one thing, understanding it can be quite another issue. The way that licensing exams are worded can be quite tricky, even for experienced plumbers. If you think you can memorize the code book and pass your test, you may be sadly mistaken. Don't stop studying the code until you understand it. Let me give you a few examples of what I'm talking about here.

The plumbing code is filled with exceptions and key words. The key words often give people the most trouble. For instance, let's assume that the code says, "All toilets shall have discharge openings which are no less than 2 inches in diameter." Now, suppose you saw this question on a true-false test, "The discharge opening of a toilet should be 2 inches in diameter." Would the answer to this test question be true or false? It would be false. The code implies that the opening must be no smaller than 2 inches. The test questions suggests that a 2-inch opening is preferable, but not mandatory. Simple little plays on words, such as I've just shown you, can ruin your test score.

The exception clauses in the code can provide as much trouble for you as the mixture of key words. For example, assume you saw a passage in the code that read like this, "All local vents serving bedpan washers, except where only one bedpan washer is being served, must be trapped." This is a simple little passage, but watch what it can do to you on a true-false test; here's the test question: "All local vents for bedpan washers must be trapped." Well, you know the answer is false, because we have just discussed it, but if you were sitting for a licensing exam, you might get pulled into giving an incorrect answer, because all vents for bedpan washers are required to be trapped, unless the vent is serving a single fixture. See how tricky things can get?

Once you have learned the code and understand what you've learned, applying it to test questions or work in the field will be much easier. I can't stress enough that you must do more than memorize the words in the code book. Take the time to learn and understand the code before you sit for your exam.

DEVELOP A STUDY PLAN

You should develop a study plan that will allow you to prepare properly for your licensing exam, without interfering too much with your daily routines. If you are already working as an apprentice for a plumbing company, you have a big advantage over someone who is working outside of the trade. Working with plumbing on a daily basis allows you to absorb a certain amount of knowledge just from association with the trade. This, however, can be more of a hindrance than a help. What do I mean by this? I'm saying that a person who works in the trade may take things for granted or be exposed to bad plumbing practices that an individual who is working at a supply warehouse wouldn't be subjected to.

People studying for a plumber's license are likely to get into trouble from their bad habits in the field. If you're working as a plumber in the field, you may be learning techniques which will influence the outcome of your licensing exam. This is not a perfect world in which we work, and not all plumbers work within the parameters of the plumbing code. I know they should, but I also know they don't. If you are learning from a plumber who is cheating the system, you can come up short on your licensing exam.

Journeyman plumbers who want to progress to the master level must often block out their day-to-day plumbing procedures in order to pass a licensing exam. Work in the field is not always done by the book, but to pass a licensing exam, the answers must come from the book. Now, I'm not talking completely about blatant abuses of the code being used in daily plumbing, but little discrepancies can result in wrong answers. For example, the plumbing inspector in your area may not require 24 hours of notice to schedule an inspection. This is fine on a daily basis, but you'd better remember that 24 hours of notice is often required to schedule an inspection.

The point to all of this is that you must shut out what you see in the field, and concentrate on what the plumbing code requires. Whether you ever go by the book completely after the day of your licensing exam or not is not the issue here. The fact is, to pass your test, you must go by the book. What happens after you are licensed is between you, your employer, and the local code enforcement office. But, for now, you must go by the book.

In order to go by the book, you must understand the book. Set aside some specific time to work on your studies. It can be first thing in the morning, your lunch break, after supper, or just before bed. When you study doesn't matter, but it does matter that you do study. If you don't set up specific times and places to study, you are less likely to stick with the program, which could result in a failed licensing exam. Determine how much time you can devote each day to studying, and make a study schedule that you can use as motivation.

CREATE A COMFORTABLE ATMOSPHERE

To make the most of your time for studying, you should create a comfortable atmosphere in which to study. If you are trying to absorb a complicated section of the code while children are playing loudly around you, there is a good chance you will not capture the true meaning of the code. Reading a code book can be difficult enough under ideal conditions, and when you put yourself in a noisy environment, your chances for success decrease.

Where should you study? The location you chose as a study area should be one in which you are in control. The exact location may be an office, a bedroom, a den, a public library, your truck, or even a travel trailer, if you happen to have one. Find a place that is quiet and which will afford you undisturbed time for studying.

GETTING INTO THE PROPER MINDSET

Getting into the proper mindset for studying can be difficult. If you're out on your own or have a family to take care of, closing off the routine worries to study can be especially hard. People who work all day and have family responsibilities may find it troublesome to set aside their other responsibilities to study the plumbing code at night and on weekends. If you have this problem, you're going to have to overcome it.

There are many different ways to gear your mind to studying. Some people like to exercise just prior to setting down to a study session. Others prefer to meditate for awhile, and still others seem to do their best studying with a radio on.

I've always found myself to be at my best late at night. Many people get tired in the evening and don't have the attention span needed to study or complete complex jobs. I, however, seem to work best during the late night and early morning hours. The phone doesn't ring, my family is asleep, and I can concentrate all of my efforts on whatever it is that I'm doing. I can't tell you what method will work best for you, but it is necessary for you to clear your mind of daily tribulations so that it is open to retaining the plumbing code.

DON'T WAIT

One key to passing your licensing exam is making sure that you don't wait until the last few days before the test to start studying. The amount of information you need to learn to pass a licensing exam is significant. If you wait until the last minute to begin studying, you are setting yourself up for failure.

I've stressed that learning the code is not enough, that you must also understand it. I know some people prefer to cram for a test just before the test date. They memorize the material for the test, and then much of it is forgotten in the weeks to follow. I don't study this way, and I don't believe the memorization method will work consistently when taking licensing exams. Due to the way that test questions are worded on licensing exams, it is often necessary to understand the question and the intent of the code fully in order to provide a correct answer.

My feelings on the memorization method of studying for a licensing exam are based on personal experience, both as an individual taking licensing exams and as an instructor of exam-preparation courses. I've conducted experiments in my code classes which have proven that tricky wording will often catch the students who cram for their exam on short notice. These same students seem to do fairly well when the test questions mirror the code, but when the questions are disguised, as they often are on real licensing exams, the students tend to receive much lower scores.

DON'T FALL INTO A ROUTINE

Don't fall into a constant routine as you study the code. If you go through your code book the same way, time and time again, you will become conditioned to give specific answers almost out of habit or reflex. This reflex motion can destroy your score on the licensing exam.

I've been a victim of falling into a routine during my studies of the code. For example, I used to know the sizing tables in the code by heart, as long as I was quoting them in the order in which I learned them. I learned the tables from the smallest pipe sizes to the largest. This allowed me to spout off the proper vent-to-trap distances and sizing requirements with ease. There was, however, a problem with this procedure.

After I had studied my code book until I thought I knew it from cover to cover, I asked a friend of mine to quiz me on the contents. As I sat, poised to provide quick answers to any questions I was asked, I suddenly realized I didn't really know the code. Oh, I knew it by heart, but only in my own way. When my friend started asking questions which were out of sequence with the way that I had studied the book, I made a lot of mistakes. Fortunately, I found my fallacies before the day of my licensing exam, and I was able to correct them.

Whether you are studying definitions, tables, charts, or piping materials, don't stay zoomed in on one topic for too long. Move around the chapters in your code book, and learn them a little here and a little there. By splitting up your study procedure, you will gain two benefits. You will not get as bored with your studies, and you will not be as likely to fall into a pattern that can be detrimental on the day of your exam.

DON'T PANIC

Don't panic when the time comes to take your licensing exam. A lot of people know their material very well, but their minds seem to go blank when they are faced with taking a test. I've never had this problem personally, but I've known a number of people who suffer from pre-test trauma. If you allow yourself to get nervous or upset just before sitting for an exam, you are dooming yourself to failure.

How can you avoid a panic attack? The best way I know for controlling pre-test jitters is to be well prepared for the test. If you know the material inside and out, you are going to pass the exam. You must, however, have confidence in your own knowledge. If there is any doubt in your mind about how proficient you are in working with the plumbing code, there will be room for fear. Eliminate the fear with proper preparation.

Taking the tests in this book is one excellent way to overcome the fear of failing your licensing exam. The questions in the chapters which follow are, in many ways, more difficult than those on actual licensing exams. The tests in this book ask questions that you are not likely to see on the real test, at least not in the volume with which I've supplied in the various chapters. If you can pass the tests in this book, you should be more than capable of passing your licensing exam.

Build your confidence by taking the tests offered in this book and by having a friend test you orally. Get your friend to thumb through your code book and ask random questions. If you can stand up to a random testing from your local code book and the tests in this book, you should have the confidence to beat the fear factor on the day of your real test.

CREATE YOUR OWN TESTS

Another way to improve your self-study of the plumbing code is to create your own tests. Feel free to use the mock tests in this book as examples of how to make your own exams. When you create your own tests, you learn the code almost by accident. In order to craft good questions, you are forced to read and understand the language in your code book. This is exactly what you should be doing as a study plan, but it comes easier when you don't realize you are studying yet. The drafting of your own mock tests will not seem like formal studying, but it can produce some of the most effective results.

After you have put together a suitable number of test questions, let them sit for several days. It is a good idea to record the proper answers on a separate piece of paper for reference after you have taken your own test. When the tests have cured for a few days, sit down and try to answer the questions correctly. Put the test answers, as you take the test, on a separate piece of paper. This will allow you to take the same tests over again as your real test date draws nearer. Since you created the test yourself, there should be very little intimidation involved with the process.

Once you have taken your own tests and scored them, you will probably be very pleased with your new knowledge. Sample tests, like the ones in this book, are one of the most effective methods you can employ to learn the code. By combining self-made tests with the tests I'm providing in this book, you should obtain excellent coverage of your local code, and your knowledge should grow quickly.

STEP AWAY FROM IT

If you study the code too intensely, you will suffer from burnout; sometimes it is best to step away from it for awhile and then go back to studying with a fresh attitude. If you force yourself to study for two hours every night, you may find yourself resenting the code and the demand on your time. If you are not receptive to learning, you will not learn. This problem can be avoided by giving yourself enough time to study in advance without having to study in a grueling manner.

DON'T GET LAZY

Don't get lazy when it comes to studying for your licensing exam. While it is fine, and even beneficial, to step away from your studies from time to time, you must discipline yourself to some study structure. It is easy to put off your studies until the next day, but there are only so many times that you can do that without coming up short. If you get lazy, you are either going to panic and cram your studies in at the last minute, or you are simply going to have a tough time passing your exam.

CHECK FOR LOCAL VARIATIONS

Before you become too comfortable with your knowledge of the code, check for local variations in the plumbing code. Regardless of what area you live or work in, the plumbing code for that area may be an amended version of the primary code. Each code jurisdiction has the right to amend and interpret the plumbing code. These amendments and interpretations can be in considerable contrast to what the code book dictates. While most areas don't make major changes in the plumbing code that has been adopted, it is well worth your time to make sure that you are studying material that will be on your licensing exam. This is especially true for individuals who will be taking exams in areas where two different codes meet or overlap.

ASK QUESTIONS

One of the best ways to learn about something new is to ask questions. This fact applies to learning the plumbing code. The code can be difficult to decipher. If there are parts of the code that you don't understand, consult a master plumber or plumbing inspector to verify your interpretation of the code. If you are already working for a plumbing company, it will be easy to ask your master plumber questions which may arise as you study for your licensing exam. If you don't have immediate access to a master plumber, record the questions you have about the code, and when you have developed a list, talk with your local plumbing inspector.

While we are on the subject of asking questions, you should ask some questions about how your licensing exam will be given. Will you be allowed to use a portable calculator? Most test sites will let you use a calculator, as long as it is not equipped with paper for printouts. Should you bring pencils to the test? Normally, testing requires the use of Number-2 pencils, which the individuals being tested are required to bring. Should you bring note paper to work with? Most test facilities will not allow people being tested to bring paper into the test center. The testing facilitators, however, will often provide paper for those who request it. When will you know the results of your licensing exam? Many test sites can give you an unofficial test score the same day that you take the exam, but the official scores are typically mailed to you at a later date. These are just some of the questions you might like to ask before sitting for your exam. The more you know about what to expect, the less you will have to fear.

Chapter 24

TROUBLESHOOTING

Plumber's Licensing Study Guide

Symptoms	Probable cause
Will not flush	No water in tank
	Stoppage in drainage system
Flushes poorly	Clogged flush holes
	Flapper or tank ball is not staying open long enough
	Not enough water in tank
	Partial drain blockage
	Defective handle
	Bad connection between handle and flush valve
	Vent is clogged
Water droplets covering tank	Condensation
Tank fills slowly	Defective ballcock
	Obstructed supply pipe
	Low water pressure
	Partially closed valve
	Partially frozen pipe
Makes unusual noises when flushed	Defective ballcock
Water runs constantly	Bad flapper or tank ball
	Bad ballcock
	Float rod needs adjusting
	Float is filled with water
	Ballcock needs adjusting
	Pitted flush valve
	Undiscovered leak
	Cracked overflow tube
Water seeps from base of toilet	Bad wax ring
	Cracked toilet bowl
Water dripping from tank	Condensation
	Bad tank-to-bowl gasket
	Bad tank-to-bowl bolts
	Cracked tank
	Flush-valve nut is loose
No water comes into the tank	Closed valve
	Defective ballcock
	Frozen pipe
	Broken pipe

FIGURE 24.1   Troubleshooting toilets.

Symptoms	Probable cause
Faucet drips from spout	Bad washers or cartridge Bad faucet seats
Faucet leaks at base of spout	Bad O ring
Faucet will not shut off	Bad washers or cartridge Bad faucet seats
Poor water pressure	Partially closed valve Clogged aerator Not enough water pressure Blockage in the faucet Partially frozen pipe
No water	Closed valve Broken pipe Frozen pipe
Drains slowly	Partial obstruction in drain or trap
Will not drain	Blocked drain or trap
Gurgles as it drains	Partial drainage blockage Partial blockage in the vent
Won't hold water	Bad basket strainer Bad putty seal on drain
Spray attachment will not spray	Clogged holes in spray head Kinked spray hose
Spray attachment will not cut off	Bad spray head

FIGURE 24.2 Troubleshooting kitchen sinks.

Symptoms	Probable cause
Faucet drips from spout	Bad washers or cartridge Bad faucet seats
Faucet leaks at base of spout	Bad O ring
Faucet will not shut off	Bad washers or cartridge Bad faucet seats
Poor water pressure	Partially closed valve Clogged aerator Not enough water pressure Blockage in the faucet Partially frozen pipe
No water	Closed valve Broken pipe Frozen pipe
Drains slowly	Partial obstruction in drain or trap
Will not drain	Blocked drain or trap
Gurgles as it drains	Partial drainage blockage Partial blockage in the vent
Won't hold water	Bad basket strainer Bad putty seal on drain

FIGURE 24.3 Troubleshooting laundry tubs.

Symptoms	Probable cause
Faucet drips from spout	Bad washers or cartridge Bad faucet seats
Faucet leaks at base of spout	Bad O ring
Faucet will not shut off	Bad washers or cartridge Bad faucet seats
Poor water pressure	Partially closed valve Clogged aerator Not enough water pressure Blockage in the faucet Partially frozen pipe
No water	Closed valve Broken pipe Frozen pipe
Drains slowly	Hair on pop-up assembly Partial obstruction in drain or trap Pop-up needs to be adjusted
Will not drain	Blocked drain or trap Pop-up is defective
Gurgles as it drains	Partial drainage blockage Partial blockage in the vent
Won't hold water	Pop-up needs adjusting Bad putty seal on drain

FIGURE 24.4 Troubleshooting lavatories.

Symptoms	Probable cause
Won't drain	Clogged drain Clogged strainer Clogged trap
Drains slowly	Hair in strainer Partial drainage blockage
Gurgles as it drains	Partial drainage blockage Partial blockage in the vent
Water drips from shower head	Bad faucet washers/cartridge Bad faucet seats
Faucet will not shut off	Bad washers or cartridge Bad faucet seats
Poor water pressure	Partially closed valve Not enough water pressure Blockage in the faucet Partially frozen pipe
No water	Closed valve Broken pipe Frozen pipe

FIGURE 24.5   Troubleshooting showers.

Symptoms	Probable cause
Relief valve leaks slowly	Bad relief valve
Relief valve blows off periodically	High water temperature High pressure in tank Bad relief valve
No hot water	Electrical power is off Elements are bad Defective thermostat Inlet valve is closed
Water not hot enough	An element is bad Bad thermostat Thermostat needs adjusting
Water too hot	Thermostat needs adjusting Controls are defective
Water leaks from tank	Hole in tank Rusted-out fitting in tank

FIGURE 24.6   Troubleshooting electric water heaters.

Symptoms	Probable cause
Wont't drain	Clogged drain Clogged tub waste Clogged trap
Drains slowly	Hair in tub waste Partial drainage blockage
Won't hold water	Tub waste needs adjusting
Won't release water	Tub waste need adjusting
Gurgles as it drains	Partial drainage blockage Partial blockage in the vent
Water drips from spout	Bad faucet washers/cartridge Bad faucet seats
Water comes out spout and shower at the same time	Bad diverter washer Bad diverter seat Bad diverter
Faucet will not shut off	Bad washers or cartridge Bad faucet seats
Poor water pressure	Partially closed valve Not enough water pressure Blockage in the faucet Partially frozen pipe
No water	Closed valve Broken pipe Frozen pipe

FIGURE 24.7   Troubleshooting bathtubs.

Symptoms	Probable cause
Relief valve leaks slowly	Bad relief valve
Relief valve blows off periodically	High water temperature High pressure in tank Bad relief valve
No hot water	Out of gas Pilot light is out Bad thermostat Control valve is off Gas valve closed
Water not hot enough	Bad thermostat Thermostat needs adjusting
Water too hot	Thermostat needs adjusting Controls are defective Burner will not shut off
Water leaks from tank	Hole in tank Rusted-out fitting in tank

FIGURE 24.8   Troubleshooting gas water heaters.

Troubleshooting Guide for Electric Water Heaters

Complaint	Cause	Solution
Water leaks (See leakage checkpoints on page 7.)	Improperly sealed hot or cold supply connections, relief valve, or thermostat threads	Tighten threaded connections
	Leakage from other appliances or water lines	Inspect other appliances near water heater
Leaking t & p valve	Thermal expansion in closed water system	Install thermal expansion tank. (do not plug t & p valve)
	Improperly seated valve	Check relief valve for proper operation
Hot water odors (Caution: unauthorized removal of the anode(s) will void the warranty. For further information, contact your dealer.)	High sulfate or mineral content in water supply	Drain and flush heater thoroughly; refill
	Bacteria in water supply	Chlorinate water supply
Not enough or no hot water	Power supply to heater is not on	Turn disconnect switch on or contact electrician
	Thermostat set too low	Refer to temperature regulation
	Heater undersized	Reduce hot water use
	Incoming water is unusually cold water (winter)	Allow more time for heater to reheat
	Leaking hot water from pipes or fixtures	Check and repair leaks
	High-temperature limit switch activated	Contact dealer to determine cause; refer to temperature regulation
Water too hot	Thermostat set too high	Refer to temperature regulation
	High-temperature limit switch activated	Contact dealer to determine cause; see temperature regulation
Water heater sounds	Scale accumulation on elements	Contact dealer to clean or replace elements
	Sediment buildup in tank bottom	Drain & flush thoroughly

SOURCE: A. O. Smith Water Products Co.

FIGURE 24.9 Troubleshooting guide for electric water heaters. *Courtesy A. O. Smith Water Products Co.*

Symptoms	Probable cause
Won't start	No electrical power Wrong voltage Bad pressure switch Bad electrical connection
Starts, but shuts off fast	Circuit breaker or fuse is inadequate Wrong voltage Bad control box Bad electrical connections Bad pressure switch Pipe blockage Pump is seized Control box is too hot
Runs, but does not produce water, or produces only a small quantity	Check valve stuck in closed position Check valve installed backward Bad electrical wiring Wrong voltage Pump is sitting above the water in the well Leak in the piping Bad pump or motor Broken pump shaft Clogged strainer Jammed impeller
Low water pressure in pressure tank	Pressure switch needs adjusting Bad pump Leak in piping Wrong voltage
Pump runs too often	Check valve stuck open Pressure tank is waterlogged and needs air injected Pressure switch needs adjusting Leak in piping Wrong-size pressure tank

FIGURE 24.10 Troubleshooting submersible potable-water pumps.

Preliminary tests—all sizes—single and three phase	
What is to be done	What it means
Measure resistance from any cable to ground (insulation resistance)	1. If the ohm value is normal, the motor windings are not grounded and the cable insulation is not damaged. 2. If the ohm value is below normal, either the windings are grounded or the cable insulation is damaged. Check the cable at the well seal as the insulation is sometimes damaged by being pinched.
Measure winding resistance (resistance between leads)	1. If all ohm values are normal, the motor windings are neither shorted nor open, and the cable colors are correct. 2. If any one ohm value is less than normal, the motor is shorted. 3. If any one ohm value is greater than normal, the winding or the cable is open, or there is a poor cable joint or connection. 4. If some ohm values are greater than normal and some less on single phase motors, the leads are mixed.

FIGURE 24.11 Troubleshooting motors.

Troubleshooting Suggestions

Causes of trouble	Motor runs continuously	
	Checking procedure	Corrective action
Pressure switch.	Switch contacts may be "welded" in closed position. Pressure switch may be set too high.	Clean contacts, replace switch, or readjust setting.
Low-level well.	Pump may exceed well capacity. Shut off pump, wait for well to recover. Check static and drawdown level from well head.	Throttle pump output or reset pump to lower level. Do not lower if sand may clog pump.
Leak in system.	Check system for leaks.	Replace damaged pipes or repair leaks.
Worn pump.	Symptoms of worn pump are similar to that of drop pipe leak or low water level in well. Reduce pressure switch setting. If pump shuts off, worn parts may be at fault. Sand is usually present in tank.	Pull pump and replace worn impellers, casing, or other close fitting parts.
Loose or broken motor shaft.	No or little water will be delivered if coupling between motor and pump shaft is loose or if a jammed pump has caused the motor shaft to shear off.	Check for damaged shafts if coupling is loose, and replace worn or defective units.
Pump screen blocked.	Restricted flow may indicate a clogged intake screen on pump. Pump may be installed in mud or sand.	Clean screen and reset at less depth. It may be necessary to clean well.
Check valve stuck closed.	No water will be delivered if check valve is in closed position.	Replace if defective.
Control box malfunction.	See pages 296, 297, and 305 for single phase.	Repair or replace.
	Motor runs but overload protector tips	
Incorrect voltage.	Using voltmeter, check the line terminals. Voltage must be within ± 10% of rated voltage.	Contact power company if voltage is incorrect.
Overheated protectors.	Direct sunlight or other heat source can make control box hot, causing protectors to trip. The box must not be hot to touch.	Shade box, provide ventilation, or move box away from heat source.
Defective control box.	For detailed procedures, see pages 295, 300, and 301.	Repair or replace.
Defective motor or cable.	For detailed procedures, see pages 294, 301, and 304.	Repair or replace.
Worn pump or motor.	Check running current. See pages 294, 305, and 306.	Replace pump and/or motor.

SOURCE: A. Y. McDonald Manufacturing Co.

FIGURE 24.12 Troubleshooting suggestions. *Courtesy A. Y. McDonald Manufacturing Co., Dubuque, Iowa.*

1. What well conditions might possibly limit the capacity of the pump?	The rate of flow from the source of supply, the diameter of a cased deep well, and the pumping level of the water in a cased deep well.
2. How does the diameter of a cased deep well and pumping level of the water affect the capacity?	They limit the size pumping equipment which can be used.
3. If there are no limiting factors, how is capacity determined?	By the maximum number of outlets or faucets likely to be in use at the same time.
4. What is suction?	A partial vacuum, created in the suction chamber of the pump, obtained by removing pressure due to atmosphere, thereby allowing greater pressure outside to force something (air, gas, water) into the container.
5. What is atmospheric pressure?	The atmosphere surrounding the earth presses against the earth and all objects on it, producing what we call atmospheric pressure.
6. How much is the pressure due to atmosphere?	This pressure varies with elevation or altitude. It is greatest at sea level (14.7 pounds per square inch) and gradually decreases as elevation above sea level is increased. The rate is approximately 1 foot per 100 feet of elevation.
7. What is maximum theoretical suction lift?	Since suction lift is actually that height to which atmospheric pressure will force water into a vacuum, theoretically we can use the maximum amount of this pressure 14.7 pounds per square inch at sea level which will raise water 33.9 feet. From this, we obtain the conversion factor of 1 pound per square inch of pressure equals 2.31-feet head.
8. How does friction loss affect suction conditions?	The resistance of the suction pipe walls to the flow of water uses up part of the work which can be done by atmospheric pressure. Therefore, the amount of loss due to friction in the suction pipe must be added to the vertical elevation which must be overcome, and the total of the two must not exceed 25 feet at sea level. This 25 feet must be reduced 1 foot for every 1000-feet elevation above sea level, which corrects for a lessened atmospheric pressure with increased elevation.
9. When and why do we use a deep-well jet pump?	The resistance of the suction pipe walls to below the pump because this is the maximum practical suction lift which can be obtained with a shallow-well pump at sea level.

FIGURE 24.13 Questions and answers about pumps. *Courtesy A. Y. McDonald Manufacturing Co., Dubuque, Iowa.*

10. What do we mean by water system?	A pump with all necessary accessories, fittings, etc., necessary for its completely automatic operation.
11. What is the purpose of a foot valve?	It is used on the end of a suction pipe to prevent the water in the system from running back into the source of supply when the pump isn't operating.
12. Name the two basic parts of a jet assembly.	Nozzle and diffuser.
13. What is the function of the nozzle?	The nozzle converts the pressure of the driving water into velocity. The velocity thus created causes a vacuum in the jet assembly or suction chamber.
14. What is the purpose of the diffuser?	The diffuser converts the velocity from the nozzle back into pressure.
15. What do we mean by "driving water"?	That water which is supplied under pressure to drive the jet.
16. What is the source of the driving water?	The driving water is continuously recirculated in a closed system.
17. What is the purpose of the centrifugal pump?	The centrifugal pump provides the energy to circulate the driving water. It also boosts the pressure of the discharged capacity.
18. Where is the jet assembly usually located in a shallow-well jet system?	Bolted to the casing of the centrifugal pump.
19. What is the principal factor which determines if a shallow-well jet system can be used?	Total suction lift.
20. When is a deep-well jet system used?	When the total suction sift exceeds that which can be overcome by atmospheric pressure.
21. Can a foot valve be omitted from a deep-well jet system? Why or why not?	No, because there are no valves in the jet assembly, and the foot valve is necessary to hold water in the system when it is primed. Also, when the centrifugal pump isn't running, the foot valve prevents the water from running back into the well.

FIGURE 24.14   Questions and answers about pumps. *Courtesy A. Y. McDonald Manufacturing Co., Dubuque, Iowa.*

22. What is the function of a check valve in the top of a submersible pump?	To hold the pressure in the line when the pump isn't running.
23. A submersible pump is made up of two basic parts. What are they?	Pump end and motor.
24. Why did the name submersible pump come into being?	Because the whole unit, pump and motor, is designed to be operated under water.
25. Can a submersible pump be installed in a 2-inch well?	No, they require a 4-inch well or larger for most domestic use. Larger pumps with larger capacities require 6-inch wells or larger.
26. A stage in a submersible pump is made up of three parts. What are they?	Impeller, diffuser, and bowl.
27. Does a submersible pump have only one pipe connection?	Yes, the discharge pipe.
28. What are two reasons we should always consider using a submersible first?	It will pump more water at higher pressure with less horsepower. It also has easier installation.
29. The amount of pressure a pump is capable of making is controlled by what?	The diameter of the impeller.
30. What do the width of an impeller and guide vane control?	The amount of water or capacity the pump is capable of pumping.

FIGURE 24.15 Questions and answers about pumps. *Courtesy A. Y. McDonald Manufacturing Co., Dubuque, Iowa.*

Normal ohm and megohm values between all leads and ground		
Insulation resistance varies very little with rating. Motors of all hp, voltage, and phase rating have similar values of insulation resistance.		
Condition of motor and leads	Ohm value	Megohm value
A new motor (without drop cable).	20,000,000 (or more)	20.0 (or more)
A used motor which can be reinstalled in the well.	10,000,000 (or more)	10.0 (or more)
Motor in well. Ohm readings are for drop cable plus motor.		
A new motor in the well.	2,000,000 (or more)	2.0 (or more)
A motor in the well in reasonably good condition.	500,000–2,000,000	0.5–2.0
A motor which may have been damaged by lightning or with damaged leads. Do not pull the pump for this reason.	20,000–500,000	0.02–0.5
A motor which definitely has been damaged or with damaged cable. The pump should be pulled and repairs made to the cable or motor replaced. The motor will not fail for this reason alone, but it will probably not operate for long.	10,000–20,000	0.01–0.02
A motor which has failed or with completely destroyed cable insulation. The pump must be pulled and the cable repaired or the motor replaced.	less than 10,000	0–0.01

FIGURE 24.16 Resistance readings.

Troubleshooting Suggestions

Cause of trouble	Checking procedure	Corrective action
Motor does not start		
No power or incorrect voltage.	Using voltmeter, check the line terminals. Voltage must be ± 10% of rated voltage.	Contact power company if voltage is incorrect.
Fuses blown or circuit fuse breakers tripped.	Check fuses for recommended size and check for loose, dirty, or corroded connections in fuse receptacle. Check for tripped circuit breaker.	Replace with proper or reset circuit breaker.
Defective pressure switch.	Check voltage at contact points. Improper contact of switch points can cause voltage less than line voltage.	Replace pressure switch or clean points.
Control box malfunction.	For detailed procedure, see ***	Repair or replace.
Defective wiring.	Check for loose or corroded connections. Check motor lead terminals with voltmeter for power.	Correct faulty wiring or connections.
Bound pump.	Locked rotor conditions can result from misalignment between pump and motor or a sand bound pump. Amp readings 3 to 6 times higher than normal will be indicated.	If pump will not start with several trials, it must be pulled and the cause corrected. New installations should always be run without turning off until water clears.
Defective cable or motor.	For detailed procedure, see pp. 291, 294, and 295.	Repair or replace.
Motor starts too often		
Pressure switch.	Check setting on pressure switch and examine for defects.	Reset limit or replace switch.
Check valve, stuck open.	Damaged or defective check valve will not hold pressure.	Replace if defective.
Waterlogged tank (air supply).	Check air-charging system for proper operation.	Clean or replace.
Leak in system.	Check system for leaks.	Replace damaged pipes or repair leaks.

SOURCE: A. Y. McDonald Manufacturing Co.

FIGURE 24.17 Troubleshooting suggestions. *Courtesy A. Y. McDonald Manufacturing Co., Dubuque, Iowa.*

Meter connections for motor testing

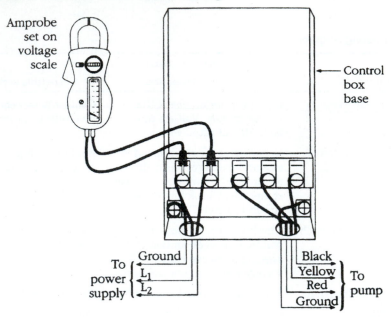

To check voltage

1. Turn power OFF

2. Remove QD cover to break all motor connections.

Caution: L1 and L2 are still connected to the power supply.

3. Turn power ON.

4. Use voltmeter as shown.

Caution: Both voltage and current tests require live circuits with power ON.

FIGURE 24.18   Meter connections for motor testing.

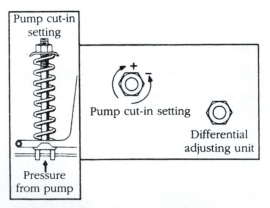

FIGURE 24.19   Fine-tuning instructions for pressure switches.

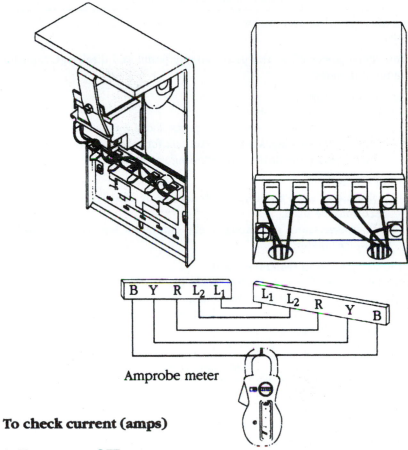

B Y R L₂ L₁   L₁ L₂ R Y B

Amprobe meter

To check current (amps)

1. Turn power OFF

2. Connect test cord as shown.

3. Turn power ON.

4. Use hook-on type ammeter as shown.

FIGURE 24.20 Checking amperage.

Single-phase control boxes

Checking and repairing procedures
(Power off)

Caution: Turn power off at the power supply panel and discharge capacitors before using ohmmeter.

A. General procedures:
  1. Disconnect line power.
  2. Inspect for damaged or burned parts, loose connections, etc.
  3. Check against diagram in control box for misconnections.
  4. Check motor insulation and winding resistance.

B. Use of ohmmeter:
  1. Ohmmeter such as Simpson Model 372 or 260. Triplet Model 630 or 666 may be used.
  2. Whenever scales are changed, clip ohmmeter lead together and "zero balance" meter.

C. Ground (insulation resistance) test:
  1. Ohmmeter Setting: Highest scale R × 10K, or R × 100K
  2. Terminal Connections: One ohmmeter lead to "Ground" terminal or Q.D. control box lid and touch other lead to the other terminals on the terminal board.
  3. Ohmmeter Reading: Pointer should remain at infinity (∞).

Additional tests

Solid state capacitor run
(CRC) control box

A. Run capacitor
  1. Meter setting: R × 1,000
  2. Connections: Red and Black leads
  3. Correct meter reading: Pointer should swing toward zero, then drift back to infinity.

B. Inductance coil
  1. Meter setting: R × 1
  2. Connections: Orange leads
  3. Correct meter reading: Less than 1 ohm.

C. Solid state switch
  Step 1 triac test
  1. Meter setting: R × 1,000
  2. Connections: R(Start) terminal and Orange lead on start switch.
  3. Correct meter reading: Should be near infinity after swing.
  Step 2 coil test
  1. Meter setting: R × 1
  2. Connections: Y(Common) and L2.
  3. Correct meter reading: Zero ohms

FIGURE 24.21 Troubleshooting motors.

> ### Single-phase control boxes
>
> Checking and repairing procedures
> (Power on)
>
> Caution: Power must be on for these tests. Do not touch any live parts.
>
> A. General procedures:
>   1. Establish line power.
>   2. Check no load voltage (pump not running).
>   3. Check load voltage (pump running).
>   4. Check current (amps) in all motor leads.
>
> B. Use of volt/amp meter:
>   1. Meter such as Amprobe Model RS300 or equivalent may be used.
>   2. Select scale for voltage or amps depending on tests.
>   3. When using amp scales, select highest scale to allow for inrush current, then select for midrange reading.
>
> C. Voltage measurements:
>   Step 1, no load.
>   1. Measure voltage at L1 and L2 of pressure switch or line contractor.
>   2. Voltage Reading: Should be ±10% of motor rating.
>   Step 2, load.
>   1. Measure voltage at load side of pressure switch or line contractor with pump running.
>   2. Voltage Reading: Should remain the same except for slight dip on starting.
>
> D. Current (amp) measurements:
>   1. Measure current on all motor leads. Use 5 conductor test cord for Q.D. control boxes.
>   2. Amp Reading: Current in Red lead should momentarily be high, then drop within one second. This verifies relay or solid state relay operation.
>
> E. Voltage symptoms:
>   1. Excessive voltage drop on starting.
>   2. Causes: Loose connections, bad contacts or ground faults, or inadequate power supply.
>
> F. Current symptoms:
>   1. Relay or switch failures will cause Red lead current to remain high and overload tripping.
>   2. Open run capacitor(s) will cause amps to be higher than normal in the Black and Yellow motor leads and lower than normal or zero amps in the Red motor lead.
>   3. Relay chatter is caused by low voltage or ground faults.
>   4. A bound pump will cause locked rotor amps and overloading tripping.
>   5. Low amps may be caused by pump running at shutoff, worn pump or stripped splines.
>   6. Failed start capacitor or open switch/relay are indicated if the red lead current is not momentarily high at starting.

FIGURE 24.22 Troubleshooting motors.

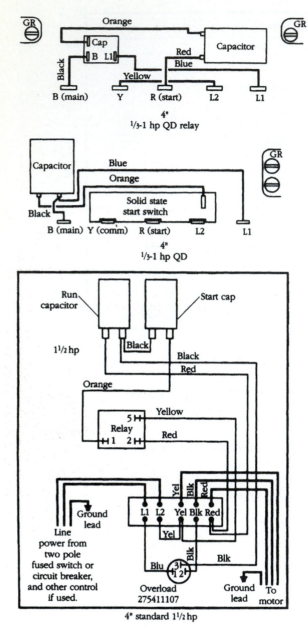

FIGURE 24.23 Wiring diagrams.

Integral horsepower control box parts					
Motor rating hp dia.	Control box (1) model no.	Part no. (2)	Capacitors MFD	Volts	Qty.
5–6"	282 2009 202	275 468 117 S	130–154	330	2
		275 479 103 (5)	15	370	2
	282 2009 203	275 468 117 S	130–154	330	2
		155 327 101 R	30	370	1
5–6" DLX	282 2009 303	275 468 117 S	130–154	330	2
		155 327 101 R	30	370	1
7½–6"	282 2019 210	275 468 119 S	270–324	330	1
		275 468 117 S	130–154	330	1
		155 327 109 R	45	370	1
	282 2019 202	275 468 117S	130–154	330	3
		275 479 103 R (5)	15	370	3
	282 2019 203	275 468 117 S	130–154	330	3
		155 327 101 R	30	370	1
		155 328 101 R	15	370	1
7½–6" DLX	282 2019 310	275 468 119 S	270–324	330	1
		275 468 117 S	130–154	330	1
		155 327 109 R	45	370	1
	282 2019 303	275 468 117 S	130–154	330	3
		155 327 101 R	30	370	1
		155 328 101 R	15	370	1
10–6"	282 2029 210	275 468 119 S	270–324	330	2
		155 327 102 R	35	370	2
	282 2029 202	275 468 117 S	130–154	330	4
		275 479 103 R (5)	15	370	5
	282 2029 203	275 468 117 S	130–154	330	4
		155 327 101 R	30	370	2
		155 328 101 R	15	370	1
	282 2029 207	275 468 119 S	270–324	330	2
		155 327 101 R	30	370	2
		155 328 101 R	15	370	1
10–6" DLX	282 2029 310	275 468 119 S	270–324	330	2
		155 327 102 R	35	370	2
	282 2029 303	275 468 117 S	130–154	330	4
		155 327 101 R	30	370	2
		155 328 101 R	15	370	1
	282 2029 307	275 468 119 S	270–324	330	2
		155 327 101 R	30	370	2
		155 328 101 R	15	370	1
15–6" DLX	282 2039 310	275 468 119 S	270–324	330	2
		155 327 109 R	45	370	3
	282 2039 303	275 468 119 S	270–324	330	2
		155 327 101 R	30	370	4
		155 328 101 R	15	370	1

FOOTNOTES:
(1) Lightning arrestor 150 814 902 suitable for all control boxes
(2) S = Start M = Main L = Line R = Run DXL = Deluxe control box with line contractor.
(3) Capacitor and overload assembly.
(4) 2 required
(5) These parts may be replaced as follows:

Old	New
275 479 102	155 328 102
275 479 103	155 328 101
275 479 105	155 328 103
275 481 102	155 327 102

FIGURE 24.24 Data chart for single-phase motors.

Integral horsepower control box parts					
Motor rating hp dia.	Control box (1) model no.	Part no.	Capacitors MFD	Volts	Qty.
1½–4"	282 3008 110	275 464 113 S	105–126	220	1
		155 328 102 R	10	370	1
	282 3007 202 or	275 461 107 S	105–126	220	1
	282 3007 102	275 479 102 R (5)	10	370	1
	282 3007 203 or	275 461 107 S	105–126	220	1
	282 3007 103	155 328 102 R	10	370	1
2–4"	282 3018 110	275 464 113 S	105–126	220	1
		155 328 103 R	20	370	1
	282 3018 202	275 464 113 S	105–126	220	1
		275 479 105 R (5)	20	370	1
	282 3018 203 or	275 464 113 S	105–126	220	1
	282 3018 103	155 328 103 R	20	370	1
2–4" DLX	282 3018 310	275 464 113 S	105–126	220	1
		155 328 103 R	20	370	1
	282 3019 103	275 464 113 S	105–126	220	1
		155 328 103 R	20	370	1
3–4"	282 3028 110	275 463 111 S	208–250	220	1
		155 327 102 R	35	370	1
	282 3028 202	275 463 111 S	208–250	220	1
		275 481 102 R (5)	35	370	2
	282 3028 203 or	275 463 111 S	208–250	220	1
	282 3028 103	155 327 102 R	35	370	1
3–4" DLX	282 3028 310	275 463 111 S	208–250	220	1
		155 327 102 R	35	370	1
	282 3029 103	275 463 111 S	208–250	220	1
		155 327 102 R	35	370	1
5–4" & 6"	282 1138 110	275 468 118 S	216–259	330	1
		155 327 101 R	30	370	2
5–4"	282 1139 202	275 468 118 S	216–259	330	1
		275 479 103 R (5)	15	370	4
	282 1139 203 or	275 468 118 S	216–259	330	1
	282 1139 003	155 327 101 R	30	370	2
5–4" & 6" DLX	282 1138 310 or	275 468 118 S	216–259	330	1
	282 1139 310	155 327 101 R	30	370	2
5–4" DLX	282 1139 303 or	275 468 118 S	216–259	330	1
	282 1139 103	155 327 101 R	30	370	2

FOOTNOTES:
(1) Lightning arrestor 150 814 902 suitable for all control boxes
(2) S = Start M = Main L = Line R = Run DLX = Deluxe control box with line contactor.
(3) Capacitor and overload assembly.
(4) 2 required
(5) These parts may be replaced as follows:

Old	New
275 479 102	155 328 102
275 479 103	155 328 101
275 479 105	155 328 103
275 481 102	155 327 102

FIGURE 24.25   Data chart for single-phase motors.

QD control box parts				
Hp	Volts	Control box model no.	(1) Solid state SW or QD (blue) relay	Start capacitor	MFD
⅓	115	2801024910	152138905(5)	275464125	159–191
		2801024915	223415905(5)	275464125	159–191
⅓	230	2801034910	152138901(5)	275464126	43–53
		2801034915	223415901(5)	275464126	43–53
½	115	2801044910	152138906(5)	275464201	250–300
		2801044915	223415906(5)	275464201	250–300
½	230	2801054910	152138902(5)	275464105	59–71
		2801054915	223415902(5)	275464105	59–71
½	230	2824055010	152138912	275470115	43–52
		2824055015	223415912(6)	275464105	59–71
¾	230	2801074910	152138903(5)	275464118	86–103
		2801074915	223415903(5)	275464118	86–103
¾	230	2824075010	152138913	275470114	108–130
		2824075015	223415913(6)	275470114	86–103
1	230	2801084910	152138904(5)	275464113	105–126
		2801084915	223415904(5)	275464113	105–126
1	230	2824085010	152138914	275470114	108–130
		2824085015	223415914(6)	275470114	108–130

FOOTNOTES:
(1) Prefixes 152 are solid state switches. Prefixes 223 are QD (Blue) Relays.
(2) Control boxes supplied with solid state relays are designed to operate on normal 230 V systems. For 208 V systems or where line voltage is between 200 V use the next larger cable size, or use boost transformer to raise the voltage to 230 V.
(3) Voltage relay kits 115 V, 305 102 901 and 230 V. 305 102 902 will replace either current voltage or QD Relays, and solid state switches.
(4) QD control boxes produced H85 or later do not contain an overload in the capacitor. On winding thermal overloads were added to three-wire motors rated ⅓-1 hp in A85. If a control box dated H85 or later is applied with a motor dated M84 or earlier, overload protection can be provided by adding an overload kit to the control box.
(5) May be replaced with QD relay kits 305 101 901 thru 906. Use same kit suffix as switch or relay suffix.
(6) Replace with CRC QD Relaying Kits, 223 415 912 with 305 105 901, 223 415 913 with 305 105 902 and 223 415 914 with 305 105 903.

FIGURE 24.26   Data chart for single-phase motors.

—NOTES—

Chapter 25
SAFETY POINTERS

General Safe Working Habits

1. Wear safety equipment.

2. Observe all safety rules at the particular location.

3. Be aware of any potential dangers in the specific situation.

4. Keep tools in good condition.

FIGURE 25.1   General safe working habits.

Safe Dressing Habits

1. Do not wear clothing that can be ignited easily.

2. Do not wear loose clothing, wide sleeves, ties or jewelry (bracelets, necklaces) that can become caught in a tool or otherwise interfere with work. This caution is especially important when working with electrical machinery.

3. Wear gloves to handle hot or cold pipes and fittings.

4. Wear heavy-duty boots. Avoid wearing sneakers on the job. Nails can easily penetrate sneakers and cause a serious injury (especially if the nail is rusty).

5. Always tighten shoelaces. Loose shoelaces can easily cause you to fall, possibly leading to injury to yourself or other workers.

6. Wear a hard hat on major construction sites to protect the head from falling objects.

FIGURE 25.2   Safe dressing habits.

Safe Operation of Grinders

1. Read the operating instructions before starting to use the grinder.

2. Do not wear any loose clothing or jewelry.

3. Wear safety glasses or goggles.

4. Do not wear gloves while using the machine.

5. Shut the machine off promptly when you are finished using it.

FIGURE 25.3   Safe operations of grinders.

Safe Use of Hand Tools

1. Use the right tool for the job.

2. Read any instructions that come with the tool unless you are thoroughly familiar with its use.

3. Wipe and clean all tools after each use. If any other cleaning is necessary, do it periodically.

4. Keep tools in good condition. Chisels should be kept sharp and any mushroomed heads kept ground smooth; saw blades should be kept sharp; pipe wrenches should be kept free of debris and the teeth kept clean; etc.

5. Do not carry small tools in your pocket, especially when working on a ladder or scaffolding. If you should fall, the tools might penetrate your body and cause serious injury.

FIGURE 25.4   Safe use of hand tools.

Safe Use of Electric Tools

1. Always use a three-prong plug with an electric tool.

2. Read all instructions concerning the use of the tool (unless you are thoroughly familiar with its use).

3. Make sure that all electrical equipment is properly grounded. Ground fault circuit interrupters (GFCI) are required by OSHA regulations in many situations.

4. Use proper-sized extension cords. (Undersized wires can burn out a motor, cause damage to the equipment, and present a hazardous situation.

5. Never run an extension cord through water or through any area where it can be cut, kinked, or run over by machinery.

6. Always hook up an extension cord to the equipment and then plug it into the main electrical outlet—not vice versa.

7. Coil up and store extension cords in a dry area.

FIGURE 25.5   Safe use of electric tools.

Rules for Working Safely in Ditches or Trenches

1. Be careful of underground utilities when digging.

2. Do not allow people to stand on the top edge of a ditch while workers are in the ditch.

3. Shore all trenches deeper than 4 feet.

4. When digging a trench, be sure to throw the dirt away from the ditch walls (2 feet or more).

5. Be careful to see that no water gets into the trench. Be especially careful in areas with a high water table. Water in a trench can easily undermine the trench walls and lead to a cave-in.

6. Never work in a trench alone.

7. Always have someone nearby—someone who can help you and locate additional help.

8. Always keep a ladder nearby so you can exit the trench quickly if need be.

9. Be watchful at all times. Be aware of any potentially dangerous situations. Remember, even heavy truck traffic nearby can cause a cave-in.

FIGURE 25.6   Working safely in ditches.

Working Safely on a Ladder

1. Use a solid and level footing to set up the ladder.

2. Use a ladder in good condition; do not use one that needs repair.

3. Be sure step ladders are opened fully and locked.

4. When using an extension ladder, place it at least ¼ of its length away from the base of the building.

5. Tie an extension ladder to the building or other support to prevent it from falling or blowing down in high winds.

6. Extend a ladder at least 3 feet over the roof line.

7. Keep both hands free when climbing a ladder.

8. Do not carry tools in your pocket when climbing a ladder. (If you fall, the tools could cut into you and cause serious injury.)

9. Use the ladder the way it should be used. For example, do not allow two people on a ladder designed for use by one person.

10. Keep the ladder and all its steps clean—free of grease, oil, mud, etc.—in order to avoid a fall and possible injury.

FIGURE 25.7   Safety on a ladder.

Safety on Rolling Scaffolds

1. Do not lay tools or other materials on the floor of the scaffold. They can easily move and you could trip over them, or they might fall, hitting someone on the ground.

2. Do not move a scaffold while you are on it.

3. Always lock the wheels when the scaffold is positioned and you are using it.

4. Always keep the scaffold level to maintain a steady platform on which to work.

5. Take no shortcuts. Be watchful at all times and be prepared for any emergencies.

FIGURE 25.8 Safety on rolling scaffolds.

To Prevent Fires

1. Always keep fire extinguishers handy, and be sure that the extinguisher is full and that you know how to use it quickly.

2. Be sure to disconnect and bleed all hoses and regulators used in welding, brazing, soldering, etc.

3. Store cylinders of acetylene, propane, oxygen, and similar substances in an upright position in a well-vented area.

4. Operate all air acetylene, welding, soldering, and related equipment according to the manufacturer's directions.

5. Do not use propane torches or other similar equipment near material that can easily catch fire.

6. Be careful at all times. Be prepared for the worst, and be ready to act.

FIGURE 25.9 Preventing fires.

—NOTES—

Appendix A

FACTS, FIGURES, CALCULATIONS AND CONVERSIONS

AVAILABLE LENGTHS OF COPPER PLUMBING TUBE

Drawn (hard copper) (feet)		*Annealed (soft copper) (feet)*	

Type K Tube

Straight Lengths:			Straight Lengths:		
Up to	8-in. diameter	20	Up to	8-in. diameter	20
	10-in. diameter	18		10-in. diameter	18
	12-in. diameter	12		12-in. diameter	12
			Coils:		
			Up to	1-in. diameter	60
					100
				1¼-in. diameter	60
					100
				2-in. diameter	40
					45

Type L Tube

Straight Lengths:			Straight Lengths:		
Up to	10-in. diameter	20	Up to	10-in. diameter	20
	12-in. diameter	18		12-inch diameter	18
			Coils:		
			Up to	1-in. diameter	60
					100
				1¼- and 1½-in. diameter	60
					100
				2-in. diameter	40
					45

DWV Tube

Straight Lengths:		Not available
All diameters	20	

Type M Tube

Straight Lengths:		Not available
All diameters	20	

FIGURE A.1   Available lengths of copper plumbing tube.

COPPER TUBE

Inside Diameter (inches)	*Nominal Size (inches)*	*Outside Diameter (inches)*
Type DWV		
N/A	1/4	0.375
N/A	3/8	0.500
N/A	1/2	0.625
N/A	5/8	0.750
N/A	3/4	0.875
N/A	1	1.125
1.295	1¼	1.375
1.511	1½	1.625
2.041	2	2.125
3.035	2½	2.625
N/A	3	3.125
N/A	3½	3.625
4.009	4	4.125
4.981	5	5.125
5.959	6	6.125
N/A	8	8.125
N/A	10	10.125
N/A	12	12.125

FIGURE A.2   Copper tube.

COPPER TUBE

Nominal Pipe Size (inches)	Outside Diameter (inches)	Inside Diameter (inches)
Type K		
1/4	0.375	0.305
3/8	0.500	0.402
1/2	0.625	0.527
5/8	0.750	0.652
3/4	0.875	0.745
1	1.125	0.995
1¼	1.375	1.245
1½	1.625	1.481
2	2.125	1.959
2½	2.625	2.435
3	3.125	2.907
3½	3.625	3.385
4	4.125	3.857
5	5.125	4.805
6	6.125	5.741
8	8.125	7.583
10	10.125	9.449
12	12.125	11.315
Type L		
1/4	0.375	0.315
3/8	0.500	0.430
1/2	0.625	0.545
5/8	0.750	0.666
3/4	0.875	0.785
1	1.125	1.025
1¼	1.375	1.265
1½	1.625	1.505
2	2.125	1.985
2½	2.625	2.465
3	3.125	2.945
3½	3.625	3.425
4	4.125	3.905
5	5.125	4.875
6	6.125	5.845
8	8.125	7.725
10	10.125	9.625
12	12.125	11.565

FIGURE A.3 Copper tube *(continued)*.

COPPER TUBE

Nominal Pipe Size (inches)	Outside Diameter (inches)	Inside Diameter (inches)
Type M		
1/4	0.375	0.325
3/8	0.500	0.450
1/2	0.625	0.569
5/8	0.750	0.690
3/4	0.875	0.811
1	1.125	1.055
1¼	1.375	1.291
1½	1.625	1.527
2	2.125	2.009
2½	2.625	2.495
3	3.125	2.981
3½	3.625	3.459
4	4.125	3.935
5	5.125	4.907
6	6.125	5.881
8	8.125	7.785
10	10.125	9.701
12	12.125	11.617

FIGURE A.3 Copper tube *(continued)*.

POLYVINYL CHLORIDE PLASTIC PIPE (PVC)

Nominal Pipe Size (inches)	Outside Diameter (inches)	Inside Diameter (inches)	Wall Thickness (inches)
1/2	0.840	0.750	0.045
3/4	1.050	0.940	0.055
1	1.315	1.195	0.060
1¼	1.660	1.520	0.070
1½	1.900	1.740	0.080
2	2.375	2.175	0.100
2½	2.875	2.635	0.120
3	3.500	3.220	0.140
4	4.500	4.110	0.195

FIGURE A.4 Polyvinyl chloride plastic pipe (PVC).

BRASS PIPE

Nominal Pipe Size (inches)	Outside Diameter (inches)	Inside Diameter (inches)	Wall Thickness (inches)
1/8	0.405	0.281	0.062
1/4	0.376	0.376	0.082
3/8	0.675	0.495	0.090
1/2	0.840	0.626	0.107
3/4	1.050	0.822	0.144
1	1.315	1.063	0.126
1¼	1.660	1.368	0.146
1½	1.900	1.600	0.150
2	2.375	2.063	0.156
2½	2.875	2.501	0.187
3	3.500	3.062	0.219

FIGURE A.5 Brass pipe.

**STEAM PIPE EXPANSION
(INCHES INCREASE PER 100 IN.)**

Temperature (°F)	Steel	Cast Iron	Brass and Copper
0	0	0	0
20	0.15	0.10	0.25
40	0.30	0.25	0.45
60	0.45	0.40	0.65
80	0.60	0.55	0.90
100	0.75	0.70	1.15
120	0.90	0.85	1.40
140	1.10	1.00	1.65
160	1.25	1.15	1.90
180	1.45	1.30	2.15
200	1.60	1.50	2.40
220	1.80	1.65	2.65
240	2.00	1.80	2.90
260	2.15	1.95	3.15
280	2.35	2.15	3.45
300	2.50	2.35	3.75
320	2.70	2.50	4.05
340	2.90	2.70	4.35
360	3.05	2.90	4.65
380	3.25	3.10	4.95
400	3.45	3.30	5.25
420	3.70	3.50	5.60
440	3.95	3.75	5.95
460	4.20	4.00	6.30
480	4.45	4.25	6.65
500	4.70	4.45	7.05

FIGURE A.6 Steam pipe expansion.

BOILING POINT OF WATER AT VARIOUS PRESSURES

Vacuum in Inches of Mercury	Boiling Point	Vacuum in Inches of Mercury	Boiling Point
29	76.62	14	181.82
28	99.93	13	184.61
27	114.22	12	187.21
26	124.77	11	189.75
25	133.22	10	192.19
24	140.31	9	194.50
23	146.45	8	196.73
22	151.87	7	198.87
21	156.75	6	200.96
20	161.19	5	202.25
19	165.24	4	204.85
18	169.00	3	206.70
17	172.51	2	208.50
16	175.80	1	210.25
15	178.91		

FIGURE A.7 Boiling point of water at various pressures.

SIZES OF ORANGEBURG[a] PIPE

Inside Diameter (inches)	Length (feet)
2	5
3	8
4	8
5	5
6	5

[a]A cardboard-type drain pipe.

FIGURE A.8 Sizes of Orangeburg® pipe.

WEIGHT OF CAST-IRON SOIL PIPE

Size (inches)	Service Weight Per Linear Foot (pounds)	Extra Heavy Size (inches)	Per Linear Foot (pounds)
2	4	2	5
3	6	3	9
4	9	4	12
5	12	5	15
6	15	6	19
7	20	8	30
8	25	10	43
		12	54
		15	75

FIGURE A.9   Weight of cast-iron soil pipe.

WEIGHT OF CAST-IRON PIPE

	Diameter (inches)	Service Weight (lb)	Extra Heavy Weight (lb)
Double Hub, 5-ft Lengths	2	21	26
	3	31	47
	4	42	63
	5	54	78
	6	68	100
	8	105	157
	10	150	225
Double Hub, 30-ft Length	2	11	14
	3	17	26
	4	23	33
Single Hub, 5-ft Lengths	2	20	25
	3	30	45
	4	40	60
	5	52	75
	6	65	95
	8	100	150
	10	145	215
Single Hub, 10-ft Lengths	2	38	43
	3	56	83
	4	75	108
	5	98	133
	6	124	160
	8	185	265
	10	270	400
No-Hub Pipe, 10-ft Lengths	1½	27	
	2	38	
	3	54	
	4	74	
	5	95	
	6	118	
	8	180	

FIGURE A.10   Weight of cast-iron pipe.

COMMON SEPTIC TANK CAPACITIES

Single-Family Dwellings; Number of Bedrooms	Multiple Dwelling Units or Apartments; One Bedroom Each	Other Uses; Maximum Fixture-Units Served	Minimum Septic Tank Capacity in Gallons
1–3		20	1000
4	2	25	1200
5–6	3	33	1500
7–8	4	45	2000
	5	55	2250
	6	60	2500
	7	70	2750
	8	80	3000
	9	90	3250
	10	100	3500

FIGURE A.11   Common septic tank capacities.

FACTS ABOUT WATER

1 ft^3 of water contains 7½ gal, 1728 in.^{3}, and weighs 62½ lb.

1 gal of water weighs 8⅓ lb and contains 231 in.^{3}

Water expands 1/23 of its volume when heated from 40° to 212°.

The height of a column of water, equal to a pressure of 1 lb/in.^{2}, is 2.31 ft.

To find the pressure in lb/in.^{2} of a column of water, multiply the height of the column in feet by 0.434.

The average pressure of the atmosphere is estimated at 14.7 lb/in.^{2} so that with a perfect vacuum it will sustain a column of water 34 ft high.

The friction of water in pipes varies as the square of the velocity.

To evaporate 1 ft^3 of water requires the consumption of 7½ lb of ordinary coal or about 1 lb of coal to 1 gal of water.

1 in.^{3} of water evaporated at atmospheric pressure is converted into approximately 1 ft^3 of steam.

FIGURE A.12   Facts about water.

RATES OF WATER FLOW

Fixture	Flow Rate (gpm)[a]
Ordinary basin faucet	2.0
Self-closing basin faucet	2.5
Sink faucet, 3/8 in.	4.5
Sink faucet, 1/2 in.	4.5
Bathtub faucet	6.0
Laundry tub cock, 1/2 in.	5.0
Shower	5.0
Ballcock for water closet	3.0
Flushometer valve for water closet	15-35
Flushometer valve for urinal	15.0
Drinking fountain	.75
Sillcock (wall hydrant)	5.0

[a]Figures do not represent the use of water-conservation devices.

FIGURE A.13   Rates of water flow.

CONSERVING WATER[a]

Activity	Normal Use (gallons)	Conservative Use (gallons)
Shower	25 (water running)	4 (wet down, soap up, rinse off)
Tub bath	36 (full)	10–12 (minimal water level)
Dishwashing	50 (tap running)	5 (wash and rinse in sink)
Toilet flushing	5–7 (depends on tank size)	1½–3 (water-saver toilets or tank displacement bottles)
Automatic dishwasher	16 (full cycle)	7 (short cycle)
Washing machine	60 (full cycle, top water level)	27 (short cycle, minimal water level)
Washing hands	2 (tap running)	1 (full basin)
Brushing teeth	1 (tap running)	1/2 (wet and rinse briefly)

[a]Amounts based on 2½–3½ gpm.

FIGURE A.14   Conserving water.

DEMAND FOR WATER AT INDIVIDUAL OUTLETS

Type of Outlet	Demand (gpm)[a]
Ordinary lavatory faucet	2.0
Self-closing lavatory faucet	2.5
Sink faucet, 3/8 in. or 1/2 in.	4.5
Sink faucet, 3/4 in.	6.0
Bath faucet, 1/2 in.	5.0
Shower head, 1/2 in.	5.0
Laundry faucet, 1/2 in.	5.0
Ballcock in water closet flush tank	3.0
1-in. flush valve (25 psi flow pressure)	35.0
1-in. flush valve (15 psi flow pressure)	27.0
3/4-in. flush valve (15 psi flow pressure)	15.0
Drinking fountain jet	0.75
Dishwashing machine (domestic)	4.0
Laundry machine (8 or 16 lb)	4.0
Aspirator (operating room or laboratory)	2.5
Hose bib or sillcock (1/2 in.)	5.0

[a]Demands do not take into account the use of water-conservation devices.

FIGURE A.15   Demand for water at individual outlets.

MAXIMUM DISTANCES FROM FIXTURE TRAPS TO VENTS

Size of Fixture (inches)	Distance from Trap to Vent[a]
1¼	2 ft 6 in.
1½	3 ft 6 in.
2	5 ft
3	6 ft
4	10 ft

[a]Figures may vary with local plumbing codes.

FIGURE A.16   Maximum distances from fixture traps to vents.

BRANCH PIPING FOR FIXTURES—MINIMUM SIZE

Fixture or Device	Size (inches)
Bathtub	1/2
Combination sink and laundry tray	1/2
Drinking fountain	3/8
Dishwashing machine (domestic)	1/2
Kitchen sink (domestic)	1/2
Kitchen sink (commercial)	3/4
Lavatory	3/8
Laundry tray (1, 2, or 3 compartments)	1/2
Shower (single head)	1/2
Sink (service, slop)	1/2
Sink (flushing rim)	3/4
Urinal (1-in. flush valve)	1
Urinal (3/4-in. flush valve)	3/4
Urinal (flush tank)	1/2
Water closet (flush tank)	3/8
Water closet (flush valve)	1
Hose bib	1/2
Wall hydrant or sillcock	1/2

FIGURE A.17   Branch piping for fixtures—minimum size.

POTENTIAL SEWAGE FLOWS

Type of Establishment	Gallons (per day per person)[a]
Schools (toilets and lavatories only)	15
Schools (with above plus cafeteria)	25
Schools (with above plus cafeteria and showers)	35
Day workers at schools and offices	15
Day camps	25
Trailer parks or tourist camps (with built-in bath)	50
Trailer parks or tourist camps (with central bathhouse)	35
Work or construction camps	50
Public picnic parks (toilet wastes only)	5
Public picnic parks (bathhouse, showers, and flush toilets)	10
Swimming pools and beaches	10
Country clubs	25 per locker
Luxury residences and estates	150
Rooming houses	40
Boarding houses	50
Hotels (with connecting baths)	50
Hotels (with private baths, 2 persons per room)	100
Boarding schools	100
Factories (gallons per person per shift, exclusive of industrial wastes)	25
Nursing homes	75
General hospitals	150
Public institutions (other than hospitals)	100
Restaurants (toilet and kitchen wastes per unit of serving capacity)	25
Kitchen wastes from hotels, camps, boarding houses, etc. that serve 3 meals per day	10
Motels	50 per bed

[a] Except for country clubs and motels.

FIGURE A.18   Potential sewage flows.

POTENTIAL SEWAGE FLOWS

Type of Establishment	Gallons
Motels with bath, toilet and kitchen wastes	60 per bed space
Drive-in theaters	5 per car space
Stores	400 per toilet room
Service stations	10 per vehicle served
Airports	3–5 per passenger
Assembly halls	2 per seat
Bowling alleys	75 per lane
Churches (small)	3–5 per sanctuary seat
Churches (large with kitchens)	5–7 per sanctuary seat
Dance halls	2 per day per person
Laundries (coin-operated)	400 per machine
Service stations	1000 (first bay) 500 (each additional bay)
Subdivisions or individual homes	75 per day per person
Marinas:	
Flush toilets	36 per fixture per hr
Urinals	10 per fixture per hr
Wash basins	15 per fixture per hr
Showers	150 per fixture per hr

FIGURE A.19   Potential sewage flows.

R VALUES FOR COMMON INSULATING MATERIALS

Material	R Value per Inch
Perlite	2.75
Loose-fill mineral wool	3.70
Extruded polystyrene	5.0
Urethane	7.2–8.0
Urethane w/foil face and 3/4-in. air space	10.0
Fiberglass batt	3.17

FIGURE A.20   R values for common insulating materials.

MELTING POINT OF VARIOUS MATERIALS

Material	Degrees Fahrenheit
Aluminum	1218
Antimony	1150
Brass	1800
Bronze	1700
Chromium	2740
Copper	2450
Gold	1975
Iron (cast)	2450
Iron (wrought)	2900
Lead	620
Manganese	2200
Mercury	39.5
Molybdenum	4500
Monel	2480
Platinum	3200
Steel (mild)	2600
Steel (stainless)	2750
Tin	450
Titanium	3360
Zinc	787

FIGURE A.21   Melting point of various materials.

WEIGHTS OF VARIOUS MATERIALS

Material	Pounds per Cubic Inch	Pounds per Cubic Foot
Aluminum	0.093	160
Antimony	0.2422	418
Brass	0.303	524
Bronze	0.320	552
Chromium	0.2348	406
Copper	0.323	558
Gold	0.6975	1205
Iron (cast)	0.260	450
Iron (wrought)	0.2834	490
Lead	0.4105	710
Manganese	0.2679	463
Mercury	0.491	849
Molybdenum	0.309	534
Monel	0.318	550
Platinum	0.818	1413
Steel (mild)	0.2816	490
Steel (stainless)	0.277	484
Tin	0.265	459
Titanium	0.1278	221
Zinc	0.258	446

FIGURE A.22 Weights of various materials.

MELTING POINTS OF COMMERCIAL METALS

Metal	Degrees Fahrenheit
Aluminum	1200
Antimony	1150
Bismuth	500
Brass	1700/1850
Copper	1940
Cadmium	610
Iron (cast)	2300
Iron (wrought)	2900
Lead	620
Mercury	139
Steel	2500
Tin	446
Zinc, cast	785

FIGURE A.23 Melting points of commercial metals.

Pipe size (in inches)	PSI	Length of pipe is 50 feet
¾	20	16
¾	40	24
¾	60	29
¾	80	34
1	20	31
1	40	44
1	60	55
1	80	65
1¼	20	84
1¼	40	121
1¼	60	151
1¼	80	177
1½	20	94
1½	40	137
1½	60	170
1½	80	200

FIGURE A.24 Discharge of pipes in gallons per minute.

Pipe size (in inches)	PSI	Length of pipe is 100 feet
¾	20	11
¾	40	16
¾	60	20
¾	80	24
1	20	21
1	40	31
1	60	38
1	80	44
1¼	20	58
1¼	40	84
1¼	60	104
1¼	80	121
1½	20	65
1½	40	94
1½	60	117
1½	80	137

FIGURE A.25 Discharge of pipes in gallons per minute.

Hose size (in inches)	Maximum outside diameter	Threads per inch
¼	1.0625	11.5

FIGURE A.26 Threads per inch for garden hose.

Valve size (in inches)	Number of turns required to operate valve
3	7.5
4	14.5
6	20.5
8	27
10	33.5

FIGURE A.27 The number of turns required to operate a double-disk valve.

Valve size (in inches)	Number of turns required to operate valve
3	11
4	14
6	20
8	27
10	33

FIGURE A.28 Number of turns required to operate a metal-seated sewerage valve.

Pipe diameter (in inches)	Approximate capacity (in U.S. gallons) per foot of pipe
¾	.0230
1	.0408
1¼	.0638
1½	.0918
2	.1632
3	.3673
4	.6528
6	1.469
8	2.611
10	4.018

FIGURE A.29　Pipe capacities.

Length (ft)	Temperature Change (°F)						
	40	50	60	70	80	90	100
20	0.278	0.348	0.418	0.487	0.557	0.626	0.696
40	0.557	0.696	0.835	0.974	1.114	1.235	1.392
60	0.835	1.044	1.253	1.462	1.670	1.879	2.088
80	1.134	1.392	1.670	1.879	2.227	2.506	2.784
100	1.192	1.740	2.088	2.436	2.784	3.132	3.480

FIGURE A.30　Thermal expansion of PVE-DVW.

Pipe material	Coefficient	
	in/in/°F	(°C)
Metallic pipe		
Carbon steel	0.000005	(14.0)
Stainless steel	0.000115	(69)
Cast iron	0.0000056	(1.0)
Copper	0.000010	(1.8)
Aluminum	0.0000980	(1.7)
Brass (yellow)	0.000001	(1.8)
Brass (red)	0.000009	(1.4)
Plastic pipe		
ABS	0.00005	(8)
PVC	0.000060	(33)
PB	0.000150	(72)
PE	0.000080	(14.4)
CPVC	0.000035	(6.3)
Styrene	0.000060	(33)
PVDF	0.000085	(14.5)
PP	0.000065	(77)
Saran	0.000038	(6.5)
CAB	0.000080	(14.4)
FRP (average)	0.000011	(1.9)
PVDF	0.000096	(15.1)
CAB	0.000085	(14.5)
HDPE	0.00011	(68)
Glass		
Borosilicate	0.0000018	(0.33)

FIGURE A.31　Thermal expansion of piping materials.

Length (ft)	Temperature Change (°F)						
	40	50	60	70	80	90	100
20	0.536	0.670	0.804	0.938	1.072	1.206	1.340
40	1.070	1.340	1.610	1.880	2.050	2.420	2.690
60	1.609	2.010	2.410	2.820	3.220	3.620	4.020
80	2.143	2.680	3.220	3.760	4.290	4.830	5.360
100	2.680	3.350	4.020	4.700	5.360	6.030	6.700

FIGURE A.32 Thermal expansion of all pipes (except PVE-DWV).

Pipe size (in inches)	Maximum outside diameter	Threads per inch
¼	1.375	8
1	1.375	8
1¼	1.6718	9
1½	1.990	9
2	2.5156	8
3	3.6239	6
4	5.0109	4
5	6.260	4
6	7.025	4

FIGURE A.33 Threads per inch for national standard thread.

Pipe size (in inches)	Maximum outside diameter	Threads per inch
¼	1.0353	14
1	1.295	11.5
1¼	1.6399	11.5
1½	1.8788	11.5
2	2.5156	8
3	3.470	8
4	4.470	8

FIGURE A.34 Threads per inch for American Standard Straight Pipe.

IPS, in	Weight per foot, lb	Length in feet containing 1 ft³ of water	Gallons in 1 linear ft
¼	0.42		0.005
⅜	0.57	754	0.0099
½	0.85	473	0.016
¾	1.13	270	0.027
1	1.67	166	0.05
1¼	2.27	96	0.07
1½	2.71	70	0.1
2	3.65	42	0.17
2½	5.8	30	0.24
3	7.5	20	0.38
4	10.8	11	0.66
5	14.6	7	1.03
6	19.0	5	1.5
8	25.5	3	2.6
10	40.5	1.8	4.1
12	53.5	1.2	5.9

FIGURE A.35 Weight of steel pipe and contained water.

Pipe	Weight factor*
Aluminum	0.35
Brass	1.12
Cast iron	0.91
Copper	1.14
Stainless steel	1.0
Carbon steel	1.0
Wrought iron	0.98

*Average plastic pipe weights one-fifth as much as carbon steel pipe.

FIGURE A.36 Relative weight factor for metal pipe.

Pipe size, in	Rod size, in
2 and smaller	⅜
2½ to 3½	½
4 and 5	⅝
6	¾
8 to 12	⅞
14 and 16	1
18	1⅛
20	1¼
24	1½

FIGURE A.37 Recommended rod size for individual pipes.

Nominal rod diameter, in	Root area of thread, in²	Maximum safe load at rod temperature of 650°F, lb
¼	0.027	240
⁵⁄₁₆	0.046	410
⅜	0.068	610
½	0.126	1,130
⅝	0.202	1,810
¾	0.302	2,710
⅞	0.419	3,770
1	0.552	4,960
1⅛	0.693	6,230
1¼	0.889	8,000
1⅜	1.053	9,470
1½	1.293	11,630
1⅝	1.515	13,630
1¾	1.744	15,690
1⅞	2.048	18,430
2	2.292	20,690
2¼	3.021	27,200
2½	3.716	33,500
2¾	4.619	41,600
3	5.621	50,600
3¼	6.720	60,500
3½	7.918	71,260

FIGURE A.38 Load rating of threaded rods.

	No water pressure	No water pressure at fixture	Low water pressure to fixture
Street water main	X		X
Curb stop	X		X
Water service	X		X
Branches		X	X
Valves	X	X	X
Stems, washers (hot and cold)		X	X
Aerator		X	X
Water meter	X	X	X

FIGURE A.39 Where to look for causes of water-pressure problems.

Nominal pipe size in inches	Outside diameter in inches	Inside diameter in inches	Weight per linear foot in pounds
¼	.375	.305	.145
⅜	.500	.402	.269
½	.625	.527	.344
⅝	.750	.652	.418
¾	.875	.745	.641
1	1.125	.995	.839
1¼	1.375	1.245	1.040
1½	1.625	1.481	1.360
2	2.125	1.959	2.060
2½	2.625	2.435	2.932
3	3.125	2.907	4.000
3½	3.625	3.385	5.122
4	4.125	3.857	6.511
5	5.125	4.805	9.672
6	6.125	5.741	13.912
8	8.125	7.583	25.900
10	10.125	9.449	40.322
12	12.125	11.315	57.802

FIGURE A.40   Facts about copper Type K tubing.

Nominal pipe size in inches	Outside diameter in inches	Inside diameter in inches	Weight per linear foot in pounds
¼	.375	.315	.126
⅜	.500	.430	.198
½	.625	.545	.285
⅝	.750	.666	.362
¾	.875	.785	.455
1	1.125	1.025	.655
1¼	1.375	1.265	.884
1½	1.625	1.505	1.111
2	2.125	1.985	1.750
2½	2.625	2.465	2.480
3	3.125	2.945	3.333
3½	3.625	3.425	4.290
4	4.125	3.905	5.382
5	5.125	4.875	7.611
6	6.125	5.845	10.201
8	8.125	7.725	19.301
10	10.125	9.625	30.060
12	12.125	11.565	40.390

FIGURE A.41   Facts about copper Type L tubing.

Nominal pipe size in inches	Outside diameter in inches	Inside diameter in inches	Weight per linear foot in pounds
¼	.375	.325	.107
⅜	.500	.450	.145
½	.625	.569	1.204
⅝	.750	.690	.263
¾	.875	.811	.328
1	1.125	1.055	.465
1¼	1.375	1.291	.682
1½	1.625	1.527	.940
2	2.125	2.009	1.460
2½	2.625	2.495	2.030
3	3.125	2.981	2.680
3½	3.625	3.459	3.580
4	4.125	3.935	4.660
5	5.125	4.907	6.660
6	6.125	5.881	8.920
8	8.125	7.785	16.480
10	10.125	9.701	25.590
12	12.125	11.617	36.710

FIGURE A.42   Facts about copper Type M tubing.

Diameter	Service weight	Extra heavy weight
2"	11	14
3"	17	26
4"	23	33

FIGURE A.43 Weight of double-hub cast-iron pipe (30-inch length).

Diameter	Service weight	Extra heavy weight
2"	38	43
3"	56	83
4"	75	108
5"	98	133
6"	124	160
8"	185	265
10"	270	400

FIGURE A.44 Weight of single-hub cast-iron pipe (10-foot length).

Size	Lbs. per ft.
2"	4
3"	6
4"	9
5"	12
6"	15
7"	20
8"	25

FIGURE A.45 Weight of cast-iron soil pipe.

Diameter	Service weight	Extra heavy weight
2"	21	26
3"	31	47
4"	42	63
5"	54	78
6"	68	100
8"	105	157
10"	150	225

FIGURE A.46 Weight of double-hub cast-iron pipe (5-foot length).

Diameter	Service weight	Extra heavy weight
2"	20	25
3"	30	45
4"	40	60
5"	52	75
6"	65	95
8"	100	150
10"	145	215

FIGURE A.47 Weight of single-hub cast-iron pipe (5-foot length).

Diameter	Weight
1½"	27
2"	38
3"	54
4"	74
5"	95
6"	118
8"	180

FIGURE A.48 Weight of no-hub cast-iron pipe (10-foot length).

Diameter	Weight
2"	5
3"	9
4"	12
5"	15
6"	19
8"	30
10"	43
12"	54
15"	75

FIGURE A.49 Weight of extra-heavy cast-iron soil pipe.

Quantity	Equals
1 square meter	10.764 square feet
	1.196 square yards
1 square centimeter	.155 square inch
1 square millimeter	.00155 square inch
.836 square meter	1 square yard
.0929 square meter	1 square foot
6.452 square centimeter	1 square inch
645.2 square millimeter	1 square inch

FIGURE A.50 Surface measures.

Quantity	Equals
144 sq. inches	1 sq. foot
9 sq. feet	1 sq. yard
1 sq. yard	1296 sq. inches
4840 sq. yards	1 acre
640 acres	1 sq. mile

FIGURE A.51 Square measure.

Quantity	Equals	Equals
7.92 inches	1 link	
100 links	1 chain	66 feet
10 chains	1 furling	660 feet
80 chains	1 mile	5280 feet

FIGURE A.52 Surveyor's measure.

Inches	Millimeters
1	25.4
2	50.8
3	76.2
4	101.6
5	127.0
6	152.4
7	177.8
8	203.2
9	228.6
10	254.0
11	279.4
12	304.8
13	330.2
14	355.6
15	381.0
16	406.4
17	431.8
18	457.2
19	482.6
20	508.0

FIGURE A.53 Inches to millimeters.

Inches^2	Millimeters^2
0.01227	8.0
0.04909	31.7
0.11045	71.3
0.19635	126.7
0.44179	285.0
0.7854	506.7
1.2272	791.7
1.7671	1140.1
3.1416	2026.8
4.9087	3166.9
7.0686	4560.4
12.566	8107.1
19.635	12667.7
28.274	18241.3
38.485	24828.9
50.265	32478.9
63.617	41043.1
78.540	50670.9

FIGURE A.54 Area in inches and millimeters.

Inches	Millimeters
0.3927	10
0.7854	20
1.1781	30
1.5708	40
2.3562	60
3.1416	80
3.9270	100
4.7124	120
6.2832	160
7.8540	200
9.4248	240
12.566	320
15.708	400
18.850	480
21.991	560
25.133	640
28.274	720
31.416	800

FIGURE A.55 Circumference in inches and millimeters.

Quantity	Unit	Symbol
Time	Second	s
Plane angle	Radius	rad
Force	Newton	N
Energy, work, quantity of heat	Joule	J
	Kilojoule	kJ
	Megajoule	MJ
Power, heat flow rate	Watt	W
	Kilowatt	kW
Pressure	Pascal	Pa
	Kilopascal	kPa
	Megapascal	MPa
Velocity, speed	Meter per second	m/s
	Kilometer per hour	km/h

FIGURE A.56 Metric symbols.

Units	Equals
1 decimeter	4 inches
1 meter	1.1 yards
1 kilometer	⅝ mile
1 hektar	2½ acres
1 stere or cu. meter	¼ cord
1 liter	1.06 qt. liquid; 0.9 qt. dry
1 hektoliter	2⅝ bushel
1 kilogram	2⅕ lbs.
1 metric ton	2200 lbs.

FIGURE A.57 Approximate metric equivalents.

Feet	Meters (m)	Millimeters (mm)
1	0.305	304.8
2	0.610	609.6
3 (1 yd.)	0.914	914.4
4	1.219	1 219.2
5	1.524	1 524.0
6 (2 yd.)	1.829	1 828.8
7	2.134	2 133.6
8	2.438	2 438.2
9 (3yd.)	2.743	2 743.2
10	3.048	3 048.0
20	6.096	6 096.0
30 (10 yd.)	9.144	9 144.0
40	12.19	12 192.0
50	15.24	15 240.0
60 (20 yd.)	18.29	18 288.0
70	21.34	21 336.0
80	24.38	24 384.0
90 (30 yd.)	27.43	27 432.0
100	30.48	30 480.0

FIGURE A.58 Length conversions.

Quantity	Unit	Symbol
Length	Millimeter	mm
	Centimeter	cm
	Meter	m
	Kilometer	km
Area	Square Millimeter	mm^7
	Square Centimeter	cm^2
	Square Decimeter	dm^2
	Square Meter	m^2
	Square Kilometer	km^2
Volume	Cubic Centimeter	cm^3
	Cubic Decimeter	dm^3
	Cubic Meter	m^3
Mass	Milligram	mg
	Gram	g
	Kilogram	kg
	Tonne	t
Temperature	Degree Celsius	°C
	Kelvin	K
Time	Second	s
Plane angle	Radius	rad
Force	Newton	N
Energy, work, quantity of heat	Joule	J
	Kilojoule	kJ
	Megajoule	MJ
Power, heat flow rate	Watt	W
	Kilowatt	kW
Pressure	Pascal	Pa
	Kilopascal	kPa
	Megapascal	MPa

FIGURE A.59   Metric symbols.

Square feet	Square meters
1	0.925
2	.1850
3	.2775
4	.3700
5	.4650
6	.5550
7	.6475
8	.7400
9	.8325
10	.9250
25	2.315
50	4.65
100	9.25

FIGURE A.60   Square feet to square meters.

Quantity	Equals
Metric cubic measure	
1000 cubic millimeters (cu. mm.)	1 cubic centimeter
1000 cubic centimeters (cu. cm.)	1 cubic decimeter
1000 cubic decimeters (cu. dm.)	1 cubic meter
Metric capacity measure	
10 milliliters (mi.)	1 centiliter
10 centiliters (cl.)	1 deciliter
10 deciliters (dl.)	1 liter
10 liters (l.)	1 dekaliter
10 dekaliters (Dl.)	1 hectoliter
10 hectoliters (hl.)	1 kiloliter
10 kiloliters (kl.)	1 myrialiter (ml.)

FIGURE A.61   Metric cubic measure.

Quantity	Equals
10 milligrams (mg.)	1 centigram
10 centigrams (cg.)	1 decigram
10 decigrams (dg.)	1 gram
10 grams (g.)	1 dekagram
10 dekagrams (Dg.)	1 hectogram
10 hectograms (hg.)	1 kilogram
10 kilogram (kg.)	1 myriagram
10 myriagrams (Mg.)	1 quintal

FIGURE A.62   Metric weight measure.

Metric linear measure		
Measure	**Equals**	**Equals**
	1 millimeter	.001 meter
10 millimeter	1 centimeter	.01 meter
10 centimeter	1 decimeter	.1 meter
10 decimeter	1 meter	1 meter
10 meters	1 dekameter	10 meters
10 dekameters	1 hectometer	100 meters
10 hectometers	1 kilometer	1000 meters
10 kilometers	1 myriameter	10,000 meters

Metric land measure	
Unit	**Equals**
1 centiare (ca.)	1 sq. meter
100 centiares (ca.)	1 are
100 ares (A.)	1 hectare
100 hectares (ha.)	1 sq. kilometer

FIGURE A.63   Metric linear measure.

1 cu. ft. at 50°F. weighs 62.41 lb.
1 gal. at 50°F weighs 8.34 lb.
1 cu. ft. of ice weighs 57.2 lb.
Water is at its greatest density at 39.2°F.
1 cu. ft. at 39.2°F. weighs 62.43 lb.

FIGURE A.64   Water weight.

Quantity	**Equals**	**Equals**
12 inches	1 foot	
3 feet	1 yard	36 inches
5½ yards	1 rod	16½ feet
40 rods	1 furlong	660 feet
8 furlongs	1 mile	5280 feet

FIGURE A.65   Linear measure.

Quantity	**Equals**
Linear measure	
12 inches	1 foot
3 feet	1 yard
5½ yards	1 rod
320 rods	1 mile
1 mile	1760 yards
1 mile	5280 feet
Square measure	
144 sq. inches	1 sq. foot
9 sq. feet	1 sq. yard
1 sq. yard	1296 sq. inches
4840 sq. yards	1 acre
640 acres	1 sq. mile

FIGURE A.66   Weights and measures.

Quantity	**Equals**
1 meter	39.3 inches
	3.28083 feet
	1.0936 yards
1 centimeter	.3937 inch
1 millimeter	.03937 inch, or nearly ⅕ inch
1 kilometer	0.62137 mile
.3048 meter	1 foot
2.54 centimeters	1 inch
	25.40 millimeters

FIGURE A.67   Metric conversions.

Quantity	**Equals**
100 sq. millimeters	1 sq. centimeter
100 sq. centimeters	1 sq. decimeter
100 sq. decimeters	1 sq. meter

FIGURE A.68   Metric square measure.

Unit	**Equals**
1 gallon	0.133681 cubic foot
1 gallon	231 cubic inches

FIGURE A.69   Volume measure equivalent.

To change	To	Multiply by
Inches	Millimeters	25.4
Feet	Meters	.3048
Miles	Kilometers	1.6093
Square inches	Square centimeters	6.4515
Square feet	Square meters	.09290
Acres	Hectares	.4047
Acres	Square kilometers	.00405
Cubic inches	Cubic centimeters	16.3872
Cubic feet	Cubic meters	.02832
Cubic yards	Cubic meters	.76452
Cubic inches	Liters	.01639
U.S. gallons	Liters	3.7854
Ounces (avoirdupois)	Grams	28.35
Pounds	Kilograms	.4536
Lbs. per sq. in. (P.S.I.)	Kg.'s per sq. cm.	.0703
Lbs. per cu. ft.	Kg.'s per cu. meter	16.0189
Tons (2000 lbs.)	Metric tons (1000 kg.)	.9072
Horsepower	Kilowatts	.746

FIGURE A.70   English to metric conversions.

Quantity	Equals
Cubic measure	
1728 cubic inches	1 cubic foot
27 cubic feet	1 cubic yard
Avoirdupois weight	
16 ounces	1 pound
100 pounds	1 hundredweight
20 hundredweight	1 ton
1 ton	2000 pounds
1 long ton	2240

FIGURE A.71   Weights and measures.

Unit	Equals
1 cu. ft.	62.4 lbs.
1 cu. ft.	7.48 gal.
1 gal.	8.33 lbs.
1 gal.	0.1337 cu. ft.

FIGURE A.72   Water volume to weight conversion.

Unit	Equals
1 meter	39.3 inches
	3.28083 feet
	1.0936 yards
1 centimeter	.3937 inch
1 millimeter	.03937 inch, or nearly ⅟₂₅ inch
1 kilometer	0.62137 mile
1 foot	.3048 meter
1 inch	2.54 centimeters
	25.40 millimeters
1 square meter	10.764 square feet
	1.196 square yards
1 square centimeter	.155 square inch
1 square millimeter	.00155 square inch
1 square yard	.836 square meter
1 square foot	.0929 square meter
1 square inch	6.452 square centimeter
	645.2 square millimeter

FIGURE A.73   Metric-customary equivalents.

Quantity	Equals
4 gills	1 pint
2 pints	1 quart
4 quarts	1 gallon
31½ gallons	1 barrel
1 gallon	231 cubic inches
7.48 gallons	1 cubic foot
1 gallon water	8.33 pounds
1 gallon gasoline	5.84 pounds

FIGURE A.74   Liquid measure.

Quantity	Equals
12 inches (in. or ")	1 foot (ft. or ')
3 feet	1 yard (yd.)
5½ yards or 16½ feet	1 rod (rd.)
40 rods	1 furlong (fur.)
8 furlongs or 320 rods	1 statute mile (mi.)

FIGURE A.75   Long measure.

Unit	Equals
1 sq. centimeter	0.1550 sq. in.
1 sq. decimeter	0.1076 sq. ft.
1 sq. meter	1.196 sq. yd.
1 are	3.954 sq. rd.
1 hektar	2.47 acres
1 sq. kilometer	0.386 sq. mile
1 sq. in.	6.452 sq. centimeters
1 sq. ft.	9.2903 sq. decimeters
1 sq. yd.	0.8361 sq. meter
1 sq. rd.	0.2529 are
1 acre	0.4047 hektar
1 sq. mile	2.59 sq. kilometers

FIGURE A.76 Square measure.

Unit	Equals
1 cubic meter	35.314 cubic feet 1.308 cubic yards 264.2 U.S. gallons (231 cubic inches)
1 cubic decimeter	61.0230 cubic inches .0353 cubic feet
1 cubic centimeter	.061 cubuic inch
1 liter	1 cubic decimeter 61.0230 cubic inches 0.0353 cubic foot 1.0567 quarts (U.S.) 0.2642 gallon (U.S.) 2.2020 lb. of water at 62°F.
1 cubic yard	.7645 cubic meter
1 cubic foot	.02832 cubic meter 28.317 cubic decimeters 28.317 liters
1 cubic inch	16.383 cubic centimeters
1 gallon (British)	4.543 liters
1 gallon (U.S.)	3.785 liters
1 gram	15.432 grains
1 kilogram	2.2046 pounds
1 metric ton	.9842 ton of 2240 pounds 19.68 cwts. 2204.6 pounds
1 grain	.0648 gram
1 ounce avoirdupois	28.35 grams
1 pound	.4536 kilograms
1 ton of 2240 lb.	1.1016 metric tons 1016 kilograms

FIGURE A.79 Measure of volume and capacity.

Volume	Weight
1 cu. ft. sand	Approx. 100 lbs.
1 cu. yd.	2700 lbs.
1 ton	¾ yd. or 20 cu. ft.
Avg. shovelful	15 lbs.
12 qt. pail	40 lbs.

FIGURE A.77 Sand volume to weight conversion.

U.S.	Metric
0.001 inch	0.025 mm
1 inch	25.400 mm
1 foot	30.48 cm
1 foot	0.3048 m
1 yard	0.9144 m
1 mile	1.609 km

FIGURE A.78 Conversion table.

U.S.	Metric
1 inch^2	6.4516 cm^2
1 feet^2	0.0929 m^2
1 yard^2	0.8361 m^2
1 acre	0.4047 ha
1 mile^2	2.590 km^2
1 inch^3	16.387 cm^3
1 feet^3	0.0283 m^3
1 yard^3	0.7647 m^3
1 U.S. ounce	29.57 ml
1 U.S. pint	0.4732 l
1 U.S. gallon	3.785 l
1 ounce	28.35 g
1 pound	0.4536 kg

FIGURE A.80 Conversion table.

	Imperial	Metric
Length	1 inch	25.4 mm
	1 foot	0.3048 m
	1 yard	0.9144 m
	1 mile	1.609 km
Mass	1 pound	0.454 kg
	1 U.S. short ton	0.9072 tonne
Area	1 ft^2	0.092 m^2
	1 yd^2	0.836 m^2
	1 acre	0.404 hectare (ha)
Capacity/Volume	1 ft^3	0.028 m^3
	1 yd^3	0.764 m^3
	1 liquid quart	0.946 litre (1)
	1 gallon	3.785 litre (1)
Heat	1 BTU	1055 joule (J)
	1 BTU/hr	0.293 watt (W)

FIGURE A.81 Measurement conversions: Imperial to metric.

To find	Multiply	By
Microns	Mils	25.4
Centimeters	Inches	2.54
Meters	Feet	0.3048
Meters	Yards	0.19144
Kilometers	Miles	1.609344
Grams	Ounces	28.349523
Kilograms	Pounds	0.4539237
Liters	Gallons (U.S.)	3.7854118
Liters	Gallons (Imperial)	4.546090
Milliliters (cc)	Fluid ounces	29.573530
Milliliters (cc)	Cubic inches	16.387064
Square centimeters	Square inches	6.4516
Square meters	Square feet	0.09290304
Square meters	Square yards	0.83612736
Cubic meters	Cubic feet	2.8316847×10^{-2}
Cubic meters	Cubic yards	0.76455486
Joules	BTU	1054.3504
Joules	Foot-pounds	1.35582
Kilowatts	BTU per minute	0.01757251
Kilowatts	Foot-pounds per minute	2.2597×10^{-5}
Kilowatts	Horsepower	0.7457
Radians	Degrees	0.017453293
Watts	BTU per minute	17.5725

FIGURE A.82 Conversion factors in converting from customary (U.S.) units to metric units.

To change	To	Multiply by
Inches	Feet	0.0833
Inches	Millimeters	25.4
Feet	Inches	12
Feet	Yards	0.3333
Yards	Feet	3
Square inches	Square feet	0.00694
Square feet	Square inches	144
Square feet	Square yards	0.11111
Square yards	Square feet	9
Cubic inches	Cubic feet	0.00058
Cubic feet	Cubic inches	1728
Cubic feet	Cubic yards	0.03703
Gallons	Cubic inches	231
Gallons	Cubic feet	0.1337
Gallons	Pounds of water	8.33
Pounds of water	Gallons	0.12004
Ounces	Pounds	0.0625
Pounds	Ounces	16
Inches of water	Pounds per square inch	0.0361
Inches of water	Inches of mercury	0.0735
Inches of water	Ounces per square inch	0.578
Inches of water	Pounds per square foot	5.2
Inches of mercury	Inches of water	13.6
Inches of mercury	Feet of water	1.1333
Inches of mercury	Pounds per square inch	0.4914
Ounces per square inch	Inches of mercury	0.127
Ounces per square inch	Inches of water	1.733
Pounds per square inch	Inches of water	27.72
Pounds per square inch	Feet of water	2.310
Pounds per square inch	Inches of mercury	2.04
Pounds per square inch	Atmospheres	0.0681
Feet of water	Pounds per square inch	0.434
Feet of water	Pounds per square foot	62.5
Feet of water	Inches of mercury	0.8824

FIGURE A.83 Measurement conversion factors.

To change	To	Multiply by
Atmospheres	Pounds per square inch	14.696
Atmospheres	Inches of mercury	29.92
Atmospheres	Feet of water	34
Long tons	Pounds	2240
Short tons	Pounds	2000
Short tons	Long tons	0.89295

FIGURE A.84 Measurement conversion factors.

Quantity	Equals	Cubic inches
2 pints	1 quart	67.2
8 quarts	1 peck	537.61
4 pecks	1 bushel	2150.42

FIGURE A.85 Dry measure.

Unit	Equals
1 gram	15.432 grains
1 kilogram	2.2046 pounds
1 metric ton	.9842 ton of 2240 pounds 19.68 cwts. 2204.6 pounds
1 grain	.0648 gram
1 ounce avoirdupois	28.35 grams
1 pound	.4536 kilograms
1 ton of 2240 lb.	1.1016 metric tons 1016 kilograms

FIGURE A.86 Weight conversions.

Inches	Millimeters
3	76.2
4	101.6
5	127
6	152.4
7	177.8
8	203.2
9	228.6
10	254

FIGURE A.87 Diameter in inches and millimeters.

Quantity	Equals	Equals
144 sq. inches	1 sq. foot	
9 sq. feeet	1 sq. yard	
30¼ sq. yards	1 sq. rod	272.25 sq. feet
160 sq. rods	1 acre	4840 sq. yards or 43,560 sq. feet
640 acres	1 sq. mile	3,097,600 sq. yards
36 sq. miles	1 township	

FIGURE A.88 Square measure.

Quantity	Equals	Meters	English equivalent
1 mm.	1 millimeter	1/1000	.03937 in.
10 mm.	1 centimeter	1/100	.3937 in.
10 cm.	1 decimeter	1/10	3.937 in.
10 dm.	1 meter	1	39.37 in.
10 m.	1 dekameter	10	32.8 ft.
10 Dm.	1 hectometer	100	328.09 ft.
10 Hm.	1 kilometer	1000	.62137 mile

FIGURE A.89 Lengths.

Barometer (ins. of mercury)	Pressure (lbs. per sq. in.)
28.00	13.75
28.25	13.88
28.50	14.00
28.75	14.12
29.00	14.24
29.25	14.37
29.50	14.49
29.75	14.61
29.921	14.696
30.00	14.74
30.25	14.86
30.50	14.98
30.75	15.10
31.00	15.23

Rule: Barometer in inches of mercury × .49116 = lbs. per sq. in.

FIGURE A.90 Atmospheric pressure per square inch.

Pipe size	Projected flow rate (gallons per minute)
½ inch	2 to 5
¾ inch	5 to 10
1 inch	10 to 20
1¼ inch	20 to 30
1½ inch	30 to 40

FIGURE A.91 Projected flow rates for various pipe sizes.

Material	Weight in pounds per cubic inch	Weight in pounds per cubic foot
Aluminum	.093	160
Antimony	.2422	418
Brass	.303	524
Bronze	.320	552
Chromium	.2348	406
Copper	.323	558
Gold	.6975	1205
Iron (cast)	.260	450
Iron (wrought)	.2834	490
Lead	.4105	710
Manganese	.2679	463
Mercury	.491	849
Molybdenum	.309	534
Monel	.318	550
Platinum	.818	1413
Steel (mild)	.2816	490
Steel (stainless)	.277	484
Tin	.265	459
Titanium	.1278	221
Zinc	.258	446

FIGURE A.93 Weights of various materials.

Pipe size	Number of gallons
¾ inch	2.8
1 inch	4.5
1¼ inch	7.8
1½ inch	11.5
2 inch	18

FIGURE A.92 Fluid volume of pipe contents for polybutylene pipe (computed on the number of gallons per 100 feet of pipe).

Pipe size	Number of gallons
1 inch	4.1
1¼ inch	6.4
1½ inch	9.2

FIGURE A.94 Fluid volume of pipe contents for copper pipe (computed on the number of gallons per 100 feet of pipe).

GPM	Liters/Minute
1	3.75
2	6.50
3	11.25
4	15.00
5	18.75
6	22.50
7	26.25
8	30.00
9	33.75
10	37.50

FIGURE A.95 Flow rate conversion from gallons per minute (GPM) to approximate liters per minute.

Inches	Decimal of an inch
1/64	0.015625
1/32	0.03125
3/64	0.046875
1/16	0.0625
5/64	0.078125
3/32	0.09375
7/64	0.109375
1/8	0.125
9/64	0.140625
5/32	0.15625
11/64	0.171875
3/16	0.1875
12/64	0.203125
7/32	0.21875
15/64	0.234375
1/4	0.25
17/64	0.265625
9/32	0.28125
19/64	0.296875
5/16	0.3125

Note: To find the decimal equivalent of a fraction, divide the numerator by the denominator.

FIGURE A.96   Decimal equivalents of fractions of and inch.

Inches	Decimal of a foot
1/8	0.01042
1/4	0.02083
3/8	0.03125
1/2	0.04167
5/8	0.05208
3/4	0.06250
7/8	0.07291
1	0.08333
1 1/8	0.09375
1 1/4	0.10417
1 3/8	0.11458
1 1/2	0.12500
1 5/8	0.13542
1 3/4	0.14583
1 7/8	0.15625
2	0.16666
2 1/8	0.17708
2 1/4	0.18750
2 3/8	0.19792
2 1/2	0.20833
2 5/8	0.21875
2 3/4	0.22917
2 7/8	0.23959
3	0.25000

Note: To change inches to decimals of a foot, divide by 12. To change decimals of a foot to inches, multiply by 12.

FIGURE A.97   Inches converted to decimals of feet.

Vacuum in inches of mercury	Boiling point
29	76.62
28	99.93
27	114.22
26	124.77
25	133.22
24	140.31
23	146.45
22	151.87
21	156.75
20	161.19
19	165.24
18	169.00
17	172.51
16	175.80
15	178.91
14	181.82
13	184.61
12	187.21
11	189.75
10	192.19
9	194.50
8	196.73
7	198.87
6	200.96
7	198.87
6	200.96
5	202.25
4	204.85
3	206.70
2	208.50
1	210.25

FIGURE A.98   Boiling points of water at various pressures.

Barometer (in. Hg)	Pressure (lb/sq in.)
28.00	13.75
28.25	13.88
28.50	14.00
28.75	14.12
29.00	14.24
29.25	14.37
29.50	14.49
29.75	14.61
29.921	14.696
30.00	14.74
30.25	14.86
30.50	14.98
30.75	15.10
31.00	15.23

Rule: Barometer in in. Hg $\times$ 0.49116 = lb/sq in.

FIGURE A.99   Atmospheric pressure per square inch.

To change	to	Multiply by
Inches	Feet	0.0833
Inches	Millimeters	25.4
Feet	Inches	12
Feet	Yards	0.3333
Yards	Feet	3
Square inches	Square feet	0.00694
Square feet	Square inches	144
Square feet	Square yards	0.11111
Square yards	Square feet	9
Cubic inches	Cubic feet	0.00058
Cubic feet	Cubic inches	1728
Cubic feet	Cubic yards	0.03703
Gallons	Cubic inches	231
Gallons	Cubic feet	0.1337
Gallons	Pounds of water	8.33
Pounds of water	Gallons	0.12004
Ounces	Pounds	0.0625
Pounds	Ounces	16
Inches of water	Pounds per square inch	0.0361
Inches of water	Inches of mercury	0.0735
Inches of water	Ounces per square inch	0.578
Inches of water	Pounds per square foot	5.2
Inches of mercury	Inches of water	13.6
Inches of mercury	Feet of water	1.1333
Inches of mercury	Pounds per square inch	0.4914
Ounces per square inch	Inches of mercury	0.127
Ounces per square inch	Inches of water	1.733
Pounds per square inch	Inches of water	27.72
Pounds per square inch	Feet of water	2.310
Pounds per square inch	Inches of mercury	2.04
Pounds per square inch	Atmospheres	0.0681
Feet of water	Pounds per square inch	0.434
Feet of water	Pounds per square foot	62.5
Feet of water	Inches of mercury	0.8824
Atmospheres	Pounds per square inch	14.696
Atmospheres	Inches of mercury	29.92
Atmospheres	Feet of water	34
Long tons	Pounds	2240
Short tons	Pounds	2000
Short tons	Long tons	0.89295

FIGURE A.100 Measurement conversion factors.

U.S.	Metric
0.001 in.	0.025 mm
1 in.	25.400 mm
1 ft	30.48 cm
1 ft	0.3048 m
1 yd	0.9144 m
1 mi	1.609 km
1 in.^{2}	6.4516 cm^2
1 ft^2	0.0929 m^2
1 yd^2	0.8361 m^2
1 a	0.4047 ha
1 mi^2	2.590 km^2
1 in.^{3}	16.387 cm^3
1 ft^3	0.0283 m^3
1 yd^3	0.7647 m^3
1 U.S. oz	29.57 ml
1 U.S. p	0.4732 l
1 U.S. gal	3.785 l
1 oz	28.35 g
1 lb	0.4536 kg

FIGURE A.101 Metric conversion table.

Grade	Ratio	Material needed for each cubic yard of concrete
Strong—watertight, exposed to weather and moderate wear	1:2¼:3	6 bags cement 14 ft^3 sand (0.52 yd^3) 18 ft^3 stone (0.67 yd^3)
Moderate—moderate strength, not exposed	1:2¾:4	5 bags cement 14 ft^3 sand (0.52 yd^3) 20 ft^3 stone (0.74 yd^3)
Economy—massive areas, low strength	1:3:5	4½ bags cement 13 ft^3 sand (0.48 yd^3) 22 ft^3 stone (0.82 yd^3)

FIGURE A.102 Formulas for concrete.

U.S.	Metric
2/12	50/300
4/12	100/300
6/12	150/300
8/12	200/300
10/12	250/300
12/12	300/300

FIGURE A.103 Roof pitches.

1 ft^3	approx. 100 lb
1 yd^3	2700 lb
1 t	¾ yd or 20 ft^3
Average shovelful	15 lb
12-qt pail	40 lb

FIGURE A.104 Volume-to-weight conversions for sand.

Minutes	Decimal of a degree	Minutes	Decimal of a degree
1	0.0166	16	0.2666
2	0.0333	17	0.2833
3	0.0500	18	0.3000
4	0.0666	19	0.3166
5	0.0833	20	0.3333
6	0.1000	21	0.3500
7	0.1166	22	0.3666
8	0.1333	23	0.3833
9	0.1500	24	0.4000
10	0.1666	25	0.4166
11	0.1833		
12	0.2000		
13	0.2166		
14	0.2333		
15	0.2500		

FIGURE A.105 Minutes converted to decimal of a degree.

−100°–30°		
°C	Base temperature	°F
−73	−100	−148
−68	−90	−130
−62	−80	−112
−57	−70	−94
−51	−60	−76
−46	−50	−58
−40	−40	−40
−34.4	−30	−22
−28.9	−20	−4
−23.3	−10	14
−17.8	0	32
−17.2	1	33.8
−16.7	2	35.6
−16.1	3	37.4
−15.6	4	39.2
−15.0	5	41.0
−14.4	6	42.8
−13.9	7	44.6
−13.3	8	46.4
−12.8	9	48.2
−12.2	10	50.0
−11.7	11	51.8
−11.1	12	53.6
−10.6	13	55.4
−10.0	14	57.2

31°–71°		
°C	Base temperature	°F
−0.6	31	87.8
0	32	89.6
0.6	33	91.4
1.1	34	93.2
1.7	35	95.0
2.2	36	96.8
2.8	37	98.6
3.3	38	100.4
3.9	39	102.2
4.4	40	104.0
5.0	41	105.8
5.6	42	107.6
6.1	43	109.4
6.7	44	111.2
7.2	45	113.0

FIGURE A.106 Temperature conversion.

31°–71°		
°C	Base temperature	°F
7.8	46	114.8
8.3	47	116.6
8.9	48	118.4
9.4	49	120.0
10.0	50	122.0
10.6	51	123.8
11.1	52	125.6
11.7	53	127.4
12.2	54	129.2
12.8	55	131.0

72°–212°		
°C	Base temperature	°F
22.2	72	161.6
22.8	73	163.4
23.3	74	165.2
23.9	75	167.0
24.4	76	168.8
25.0	77	170.6
25.6	78	172.4
26.1	79	174.2
26.7	80	176.0
27.8	81	177.8
28.3	82	179.6
28.9	83	181.4
29.4	84	183.2
30.0	85	185.0
30.6	86	186.8
31.1	87	188.6
31.7	88	190.4
32.2	89	192.2
32.8	90	194.0
33.3	91	195.8
33.9	92	197.6
34.4	93	199.4
35.0	94	201.2
35.6	95	203.0
36.1	96	204.8

FIGURE A.106 Temperature conversion *(continued)*.

213°–620°		
°C	Base temperature	°F
104	220	248
110	230	446
116	240	464
121	250	482
127	260	500
132	270	518
138	280	536
143	290	554
149	300	572
154	310	590
160	320	608
166	330	626
171	340	644
177	350	662
182	360	680

FIGURE A.106 Temperature conversion *(continued)*.

213°–620°		
°C	Base temperature	°F
188	370	698
193	380	716
199	390	734
204	400	752
210	410	770
216	420	788
221	430	806
227	440	824
232	450	842
238	460	860

621°–1000°		
°C	Base temperature	°F
332	630	1166
338	640	1184
343	650	1202
349	660	1220
354	670	1238
360	680	1256
366	690	1274
371	700	1292
377	710	1310
382	720	1328
388	730	1346
393	740	1364
399	750	1382
404	760	1400
410	770	1418
416	780	1436
421	790	1454
427	800	1472
432	810	1490
438	820	1508
443	830	1526
449	840	1544
454	850	1562
460	860	1580
466	870	1598

FIGURE A.106 Temperature conversion *(continued)*.

To change	to	Multiply by
Inches	Feet	0.0833
Inches	Millimeters	25.4
Feet	Inches	12
Feet	Yards	0.3333
Yards	Feet	3
Square inches	Square feet	0.00694
Square feet	Square inches	144
Square feet	Square yards	0.11111
Square yards	Square feet	9
Cubic inches	Cubic feet	0.00058
Cubic feet	Cubic inches	1728
Cubic feet	Cubic yards	0.03703
Cubic yards	Cubic feet	27
Cubic inches	Gallons	0.00433
Cubic feet	Gallons	7.48
Gallons	Cubic inches	231
Gallons	Cubic feet	0.1337
Gallons	Pounds of water	8.33
Pounds of water	Gallons	0.12004
Ounces	Pounds	0.0625
Pounds	Ounces	16
Inches of water	Pounds per square inch	0.0361
Inches of water	Inches of mercury	0.0735
Inches of water	Ounces per square inch	0.578
Inches of water	Pounds per square foot	5.2
Inches of mercury	Inches of water	13.6
Inches of mercury	Feet of water	1.1333
Inches of mercury	Feet of water	0.4914
Ounces per square inch	Pounds per square inch	0.127
Ounces per square inch	Inches of mercury	1.733
Pounds per square inch	Inches of water	27.72
Pounds per square inch	Feet of water	2.310
Pounds per square inch	Inches of mercury	2.04
Pounds per square inch	Atmospheres	0.0681
Feet of water	Pounds per square inch	0.434
Feet of water	Pounds per square foot	62.5
Feet of water	Inches of mercury	0.8824
Atmospheres	Pounds per square inch	14.696
Atmospheres	Inches of mercury	29.92
Atmospheres	Feet of water	34
Long tons	Pounds	2240
Short tons	Pounds	2000
Short tons	Long tons	0.89295

FIGURE A.107 Useful multipliers.

Ounces	Kilograms
1	0.028
2	0.057
3	0.085
4	0.113
5	0.142
6	0.170
7	0.198
8	0.227
9	0.255
10	0.283
11	0.312
12	0.340
13	0.369
14	0.397
15	0.425
16	0.454

FIGURE A.108 Ounces to kilograms.

Pounds	Kilograms
1	0.454
2	0.907
3	1.361
4	1.814
5	2.268
6	2.722
7	3.175
8	3.629
9	4.082
10	4.536
25	11.34
50	22.68
75	34.02
100	45.36

FIGURE A.109 Pounds to kilograms.

Gallons per minute	Liters per minute
1	3.75
2	6.50
3	11.25
4	15.00
5	18.75
6	22.50
7	26.25
8	30.00
9	33.75
10	37.50

FIGURE A.110 Flow-rate conversion.

1 GMP 0.134 cu ft/min	
1 cu ft/min (cfm) 448.8 gal/hr (gph)	
Feet per second	Meters per second
1	0.3050
2	0.610
3	0.915
4	1.220
5	1.525
6	1.830
7	2.135
8	2.440
9	2.754
10	3.050

FIGURE A.111 Flow-rate equivalents.

Pounds per square foot	Kilopascals
1	0.0479
2	0.0958
3	0.1437
4	0.1916
5	0.2395
6	0.2874
7	0.3353
8	0.3832
9	0.4311
10	0.4788
25	1.1971
50	2.394
75	3.5911
100	4.7880

FIGURE A.112 Pounds per square foot to kilo-pascals.

Pounds per square inch	Kilopascals
1	6.895
2	13.790
3	20.685
4	27.580
5	34.475
6	41.370
7	48.265
8	55.160
9	62.055
10	68.950
25	172.375
50	344.750
75	517.125
100	689.500

FIGURE A.113 Pounds per square inch to kilo-pascals.

Feet head	Pounds per square inch	Feet head	Pounds per square inch
1	0.43	50	21.65
2	0.87	60	25.99
3	1.30	70	30.32
4	1.73	80	34.65
5	2.17	90	38.98
6	2.60	100	43.34
7	3.03	110	47.64
8	3.46	120	51.97
9	3.90	130	56.30
10	4.33	140	60.63
15	6.50	150	64.96
20	8.66	160	69.29
25	10.83	170	73.63
30	12.99	180	77.96
40	17.32	200	86.62

FIGURE A.114 Water feet head to pounds per square inch.

1/32	0.03125
1/16	0.0625
3/32	0.09375
1/8	0.125
5/32	0.15625
3/16	0.1875
7/32	0.21875
1/4	0.25
9/32	0.28125
5/16	0.3125
11/32	0.34375
3/8	0.375
13/32	0.40625
7/16	0.4375
15/32	0.46875
1/2	0.500
17/32	0.53125
9/16	0.5625
19/32	0.59375
5/8	0.625
21/32	0.65625
11/16	0.6875
23/32	0.71875
3/4	0.75
25/32	0.78125
13/16	0.8125
27/32	0.84375
7/8	0.875
29/32	0.90625
15/16	0.9375
31/32	0.96875
1	1.000

FIGURE A.115 Decimal equivalents of an inch.

Pounds per square inch	Feet head
1	2.31
2	4.62
3	6.93
4	9.24
5	11.54
6	13.85
7	16.16
8	18.47
9	20.78
10	23.09
15	34.63
20	46.18
25	57.72
30	69.27
40	92.36
50	115.45
60	138.54
70	161.63
80	184.72
90	207.81
100	230.90
110	253.98
120	277.07
130	300.16
140	323.25
150	346.34
160	369.43
170	392.52
180	415.61
200	461.78
250	577.24
300	692.69
350	808.13
400	922.58
500	1154.48
600	1385.39
700	1616.30
800	1847.20
900	2078.10
1000	2309.00

FIGURE A.116 Water pressure in pounds with equivalent feet heads.

Outside design temperature	= average of lowest recorded temperature in each month from October to March
Inside design temperature	= 70°F or as specified by owner

A degree day is one day multiplied by the **number of Fahrenheit degrees** the mean temperature is below 65°F. The number of **degree days** in a year is a good guideline for designing heating and insulation systems.

FIGURE A.117   Design temperature.

To find	Multiply	By
Microns	Mils	25.4
Centimeters	Inches	2.54
Meters	Feet	0.3048
Meters	Yards	0.19144
Kilometers	Miles	1.609344
Grams	Ounces	28.349523
Kilograms	Pounds	0.4539237
Liters	Gallons (U.S.)	3.7854118
Liters	Gallons (imperial)	4.546090
Milliliters (cc)	Fluid ounces	29.573530
Milliliters (cc)	Cubic inches	16.387064
Square centimeters	Square inches	6.4516
Square meters	Square feet	0.09290304
Square meters	Square yards	0.83612736
Cubic meters	Cubic feet	2.8316847×10^{-2}
Cubic meters	Cubic yards	0.76455486
Joules	Btu	1054.3504
Joules	Foot-pounds	1.35582
Kilowatts	Btu per minute	0.01757251
Kilowatts	Foot-pounds per minute	2.2597×10^{-5}
Kilowatts	Horsepower	0.7457
Radians	Degrees	0.017453293
Watts	Btu per minute	17,5725

FIGURE A.118   Factors used in converting from customary (U.S.) units to metric units.

Horsepower	Kilowatt
1/20	0.025
1/16	0.05
1/8	0.1
1/6	0.14
1/4	0.2
1/3	0.28
1/2	0.4
1	0.8
1½	1.1

FIGURE A.119   Metric motor ratings.

1 ft³	62.4 lbs
1 ft³	7.48 gal
1 gal	8.33 lbs
1 gal	0.1337 ft³

FIGURE A.120   Water volume to weight.

1 ft³ at 50°F weighs 62.41 lb
1 gal at 50°F weighs 8.34 lb
1 ft³ of ice weighs 57.2 lb
Water is at its greatest density at 39.2°F
1 ft³ at 39.2°F weighs 62.43 lb

FIGURE A.121   Water weight.

Unit	Symbol
Length	
Millimeter	mm
Centimeter	cm
Meter	m
Kilometer	km
Area	
Square millimeter	mm^2
Square centimeter	cm^2
Square decimeter	dm^2
Square meter	m^2
Square kilometer	km^2
Volume	
Cubic centimeter	cm^3
Cubic decimeter	dm^3
Cubic meter	m^3
Mass	
Milligram	mg
Gram	g
Kilogram	kg
Tonne	t
Temperature	
Degrees Celsius	°C
Kelvin	K
Time	
Second	s
Plane angle	
Radius	rad
Force	
Newton	N
Energy, work, quantity of heat	
Joule	J
Kilojoule	kJ
Megajoule	MJ
Power, heat flow rate	
Watt	W
Kilowatt	kW
Pressure	
Pascal	Pa
Kilopascal	kPa
Megapascal	MPa
Velocity, speed	
Meter per second	m/s
Kilometer per hour	km/h

FIGURE A.122   Metric abbreviations.

Watt output	Millihorsepower (MHP)	Fractional HP
0.746	1	1/1000
1.492	2	1/500
2.94	4	1/250
4.48	6	1/170
5.97	8	1/125
7.46	10	1/100
9.33	12.5	1/80
10.68	14.3	1/70
11.19	15	1/65
11.94	16	1/60
14.92	20	1/50
18.65	25	1/40
22.38	30	1/35
24.90	33	1/30

FIGURE A.124   Motor power output comparison.

Length	
1 in.	25.4 mm
1 ft	0.3048 m
1 yd	0.9144 m
1 mi	1.609 km
Mass	
1 lb	0.454 kb
1 U.S. short ton	0.9072 t
Area	
1 ft^2	0.092 m^2
1 yd^2	0.836 m^2
1 a	0.404 ha
Capacity, liquid	
1 ft^3	0.028 m^3
Capacity, dry	
1 yd^3	0.764 m^3
Volume, liquid	
1 qt	0.946 l
1 gal	3.785 l
Heat	
1 Btu	1055 joule (J)
1 Btu/hr	0.293 watt (W)

FIGURE A.125   Measurements conversions (imperial to metric).

Inch scale		Metric scale
¼	is closest to	1:50
⅛	is closest to	1:100

FIGURE A.123   Scales used for building plans.

Inch scale		Metric scale
1/16	is closest to	1:200

FIGURE A.126   Scale used for site plans.

```
16 oz = 1 lb
100 lb = 1 cwt
20 cwt = 1 ton
1 ton = 2000 lb
1 long ton = 2240 lb

16 dr = 1 oz
16 oz = 1 lb
100 lb = 1 cwt
20 cwt = 1 ton = 2000 lb
```

FIGURE A.127 Avoirdupois weight.

10 milligrams (mg)	1 centigram
10 centigrams (cg)	1 decigram
10 decigrams (dg)	1 gram
10 grams (g)	1 dekagram
10 dekagrams (dkg)	1 hectogram
10 hectograms (hg)	1 kilogram
10 kilograms (kg)	1 myriagram
10 myriagrams (myg)	1 quintal

FIGURE A.128 Metric weight measure.

60 s	1 min
60 min	1°
360°	1 circle

FIGURE A.129 Circular measure.

1 gal	0.133681 ft^3
1 gal	231 in.^{3}

FIGURE A.130 Volume measure equivalents.

Temperature (°F)	Steel	Cast iron	Brass and copper
0	0	0	0
20	0.15	0.10	0.25
40	0.30	0.25	0.45
60	0.45	0.40	0.65
80	0.60	0.55	0.90
100	0.75	0.70	1.15
120	0.90	0.85	1.40
140	1.10	1.00	1.65
160	1.25	1.15	1.90
180	1.45	1.30	2.15
200	1.60	1.50	2.40
220	1.80	1.65	2.65
240	2.00	1.80	2.90
260	2.15	1.95	3.15
280	2.35	2.15	3.45
300	2.50	2.35	3.75
320	2.70	2.50	4.05
340	2.90	2.70	4.35
360	3.05	2.90	4.65
380	3.25	3.10	4.95
400	3.45	3.30	5.25
420	3.70	3.50	5.60
440	3.95	3.75	5.95
460	4.20	4.00	6.30
480	4.45	4.25	6.65
500	4.70	4.45	7.05

FIGURE A.131 Steam pipe expansion (inches increase per 100 inches).

Single-family dwellings; number of bedrooms	Multiple dwelling units or apartments; one bedroom each	Other uses; maximum fixture-units served	Minimum septic tank capacity in gallons
1–3		20	1000
4	2	25	1200
5–6	3	33	1500
7–8	4	45	2000
	5	55	2250
	6	60	2500
	7	70	2750
	8	80	3000
	9	90	3250
	10	100	3500

FIGURE A.132   Common septic tank capacities.

Appliance	Size (inches)
Clothes washer	2
Bathtub with or without shower	1½
Bidet	1½
Dental unit or cuspidor	1¼
Drinking fountain	1¼
Dishwasher, domestic	1½
Dishwasher, commercial	2
Floor drain	2, 3, or 4
Lavatory	1¼
Laundry tray	1½
Shower stall, domestic	2
Sinks:	
Combination, sink and tray (with disposal unit)	1½
Combination, sink and tray (with one trap)	1½
Domestic, with or without disposal unit	1½
Surgeon's	1½
Laboratory	1½
Flushrim or bedpan washer	3
Service sink	2 or 3
Pot or scullery sink	2
Soda fountain	1½
Commercial, flat rim, bar, or counter sink	1½
Wash sinks circular or multiple	1½
Urinals:	
Pedestal	3
Wall-hung	1½ or 2
Trough (per 6-ft section)	1½
Stall	2
Water closet	3

FIGURE A.133   Common trap sizes.

Category	Estimated water usage per day
Barber shop	100 gal per chair
Beauty shop	125 gal per chair
Boarding school, elementary	75 gal per student
Boarding school, secondary	100 gal per student
Clubs, civic	3 gal per person
Clubs, country	25 gal per person
College, day students	25 gal per student
College, junior	100 gal per student
College, senior	100 gal per student
Dentist's office	750 gal per chair
Department store	40 gal per employee
Drugstore	500 gal per store
Drugstore with fountain	2000 gal per store
Elementary school	16 gal per student
Hospital	400 gal per patient
Industrial plant	30 gal per employee + process water
Junior and senior high school	25 gal per student
Laundry	2000–20,000 gal
Launderette	1000 gal per unit
Meat market	5 gal per 100 ft^2 of floor area
Motel or hotel	125 gal per room
Nursing home	150 gal per patient
Office building	25 gal per employee
Physician's office	200 gal per examining room
Prison	60 gal per inmate
Restaurant	20–120 gal per seat
Rooming house	100 gal per tenant
Service station	600–1500 gal per stall
Summer camp	60 gal per person
Theater	3 gal per seat

FIGURE A.134   Estimating guidelines for daily water usage.

Unit	Pounds per square inch	Feet of water	Meters of water	Inches of mercury	Atmospheres
1 pound per square inch	1.0	2.31	0.704	2.04	0.0681
1 foot of water	0.433	1.0	0.305	0.882	0.02947
1 meter of water	1.421	3.28	1.00	2.89	0.0967
1 inch of mercury	0.491	1.134	0.3456	1.00	0.0334
1 atmosphere (sea level)	14.70	33.93	10.34	29.92	1.0000

FIGURE A.135   Conversion of water values.

To convert	Multiply by	To obtain
A		
acres	4.35×10^4	sq. ft.
acres	4.047×10^3	sq. meters
acre-feet	4.356×10^4	cu. feet
acre-feet	3.259×10^5	gallons
atmospheres	2.992×10^1	in. of mercury (at 0°C.)
atmospheres	1.0333	kgs./sq. cm.
atmospheres	1.0333×10^4	kgs./sq. meter
atmospheres	1.47×10^1	pounds/sq. in.
B		
barrels (u.s., liquid)	3.15×10^1	gallons
barrels (oil)	4.2×10^1	gallons (oil)
bars	9.869×10^{-1}	atmospheres
btu	7.7816×10^2	foot-pounds
btu	3.927×10^{-4}	horsepower-hours
btu	2.52×10^{-1}	kilogram-calories
btu	2.928×10^{-4}	kilowatt-hours
btu/hr.	2.162×10^{-1}	ft. pounds/sec.
btu/hr.	3.929×10^{-4}	horsepower
btu/hr.	2.931×10^{-1}	watts
btu/min.	1.296×10^1	ft.-pounds/sec.
btu/min.	1.757×10^{-2}	kilowatts
C		
centigrade (degrees)	(°C × 9/5) + 32	fahrenheit (degrees)
centigrade (degrees)	°C + 273.18	kelvin (degrees)
centigrams	$1. \times 10^{-2}$	grams
centimeters	3.281×10^{-2}	feet
centimeters	3.937×10^{-1}	inches
centimeters	$1. \times 10^{-5}$	kilometers
centimeters	$1. \times 10^{-2}$	meters
centimeters	$1. \times 10^1$	millimeters
centimeters	3.937×10^2	mils
centimeters of mercury	1.316×10^{-2}	atmospheres
centimeters of mercury	4.461×10^{-1}	ft. of water
centimeters of mercury	1.934×10^{-1}	pounds/sq. in.
centimeters/sec.	1.969	feet/min.
centimeters/sec.	3.281×10^{-2}	feet/sec.
centimeters/sec.	6.0×10^{-1}	meters/min.
centimeters/sec./sec.	3.281×10^{-2}	ft./sec./sec.
cubic centimeters	3.531×10^{-5}	cubic ft.
cubic centimeters	6.102×10^{-2}	cubic in.
cubic centimeters	1.0×10^{-6}	cubic meters
cubic centimeters	2.642×10^{-4}	gallons (u.s. liquid)
cubic centimeters	2.113×10^{-3}	pints (u.s. liquid)
cubic centimeters	1.057×10^{-3}	quarts (u.s. liquid)

FIGURE A.136 Measurement conversions.

To convert	Multiply by	To obtain
cubic feet	2.8320×10^4	cu. cms.
cubic feet	1.728×10^3	cu. inches
cubic feet	2.832×10^{-2}	cu. meters
cubic feet	7.48052	gallons (u.s. liquid)
cubic feet	5.984×10^1	pints (u.s. liquid)
cubic feet	2.992×10^1	quarts (u.s. liquid)
cubic feet/min.	4.72×10^1	cu. cms./sec.
cubic feet/min.	1.247×10^{-1}	gallons/sec.
cubic feet/min.	4.720×10^{-1}	liters/sec.
cubic feet/min.	6.243×10^1	pounds water/min.
cubic feet/sec.	6.46317×10^{-1}	million gals./day
cubic feet/sec.	4.48831×10^2	gallons/min.
cubic inches	5.787×10^{-4}	cu. ft.
cubic inches	1.639×10^{-5}	cu. meters
cubic inches	2.143×10^{-5}	cu. yards
cubic inches	4.329×10^{-3}	gallons

D

To convert	Multiply by	To obtain
degrees (angle)	1.745×10^{-2}	radians
degrees (angle)	3.6×10^3	seconds
degrees/sec.	2.778×10^{-3}	revolutions/sec.
dynes/sq. cm.	4.015×10^{-4}	in. of water (at 4°C.)
dynes	1.020×10^{-6}	kilograms
dynes	2.248×10^{-6}	pounds

F

To convert	Multiply by	To obtain
fathoms	1.8288	meters
fathoms	6.0	feet
feet	3.048×10^1	centimeters
feet	3.048×10^{-1}	meters
feet of water	2.95×10^{-2}	atmospheres
feet of water	3.048×10^{-2}	kgs./sq. cm.
feet of water	6.243×10^1	pounds/sq. ft.
feet/min.	5.080×10^{-1}	cms./sec.
feet/min.	1.667×10^{-2}	feet/sec.
feet/min.	3.048×10^{-1}	meters/min.
feet/min.	1.136×10^{-2}	miles/hr.
feet/sec.	1.829×10^1	meters/min.
feet/100 feet	1.0	per cent grade
foot-pounds	1.286×10^{-3}	btu
foot-pounds	1.356×10^7	ergs
foot-pounds	3.766×10^{-7}	kilowatt-hrs.
foot-pounds/min.	1.286×10^{-3}	btu/min.
foot-pounds/min.	3.030×10^{-5}	horsepower
foot-pounds/min.	3.241×10^{-4}	kg.-calories/min.
foot-pounds/sec.	4.6263	btu/hr.
foot-pounds/sec.	7.717×10^{-2}	btu/min.
foot-pounds/sec.	1.818×10^{-3}	horsepower
foot-pounds/sec.	1.356×10^{-3}	kilowatts
furlongs	1.25×10^{-1}	miles (u.s.)

FIGURE A.136 Measurement conversions *(continued)*.

To convert	Multiply by	To obtain
	G	
gallons	3.785×10^3	cu. cms.
gallons	1.337×10^{-1}	cu. feet
gallons	2.31×10^2	cu. inches
gallons	3.785×10^{-3}	cu. meters
gallons	4.951×10^{-3}	cu. yards
gallons	3.785	liters
gallons (liq. br. imp.)	1.20095	gallons (u.s. liquid)
gallons (u.s.)	$8,3267 \times 10^{-1}$	gallons (imp.)
gallons of water	8.337	pounds of water
gallons/min.	2.228×10^{-3}	cu. feet/sec.
gallons/min.	6.308×10^{-2}	liters/sec.
gallons/min.	8.0208	cu. feet/hr.
	H	
horsepower	4.244×10^1	btu/min.
horsepower	3.3×10^4	foot-lbs./min.
horsepower	5.50×10^2	foot-lbs./sec.
horsepower (metric)	9.863×10^{-1}	horsepower
horsepower	1.014	horsepower (metric)
horsepower	7.457×10^{-1}	kilowatts
horsepower	7.457×10^2	watts
horsepower (boiler)	3.352×10^4	btu/hr.
horsepower (boiler)	9.803	kilowatts
horsepower-hours	2.547×10^3	btu
horsepower-hours	1.98×10^6	foot-lbs.
horsepower-hours	6.4119×10^5	gram-calories
hours	5.952×10^{-3}	weeks
	I	
inches	2.540	centimeters
inches	2.540×10^{-2}	meters
inches	1.578×10^{-5}	miles
inches	2.54×10^1	millimeters
inches	1.0×10^3	mils
inches	2.778×10^{-2}	yards
inches of mercury	3.342×10^{-2}	atmospheres
inches of mercury	1.133	feet of water
inches of mercury	3.453×10^{-2}	kgs./sq. cm.
inches of mercury	3.453×10^2	kgs./sq. meter
inches of mercury	7.073×10^1	pounds/sq. ft.
inches of mercury	4.912×10^{-1}	pounds/sq. in.
in. of water (at 4°C.)	7.355×10^{-2}	inches of mercury
in. of water (at 4°C.)	2.54×10^{-3}	kgs./sq. cm.
in. of water (at 4°C.)	5.204	pounds/sq. ft.
in. of water (at 4°C.)	3.613×10^{-2}	pounds/sq. in.

FIGURE A.136 Measurement conversions *(continued).*

To convert	Multiply by	To obtain
	J	
joules	9.486×10^{-4}	btu
joules/cm.	1.0×10^{7}	dynes
joules/cm.	1.0×10^{2}	joules/meter (newtons)
joules/cm.	2.248×10^{1}	pounds
	K	
kilograms	9.80665×10^{5}	dynes
kilograms	1.0×10^{3}	grams
kilograms	2.2046	pounds
kilograms	9.842×10^{-4}	tons (long)
kilograms	1.102×10^{-3}	tons (short)
kilograms/sq. cm.	9.678×10^{-1}	atmospheres
kilograms/sq. cm.	3.281×10^{1}	feet of water
kilograms/sq. cm.	2.896×10^{1}	inches of mercury
kilograms/sq. cm.	1.422×10^{1}	pounds/sq. in.
kilometers	1.0×10^{5}	centimeters
kilometers	3.281×10^{3}	feet
kilometers	3.937×10^{4}	inches
kilometers	1.0×10^{3}	meters
kilometers	6.214×10^{-1}	miles (statute)
kilometers	5.396×10^{-1}	miles (nautical)
kilometers	1.0×10^{6}	millimeters
kilowatts	5.692×10^{1}	btu/min.
kilowatts	4.426×10^{4}	foot-lbs./min.
kilowatts	7.376×10^{2}	foot-lbs./sec.
kilowatts	1.341	horsepower
kilowatts	1.434×10^{1}	kg.-calories/min.
kilowatts	1.0×10^{3}	watts
kilowatt-hrs.	3.413×10^{3}	btu
kilowatt-hrs.	2.655×10^{6}	foot-lbs.
kilowatt-hrs.	8.5985×10^{3}	gram calories
kilowatt-hrs.	1.341	horsepower-hours
kilowatt-hrs.	3.6×10^{6}	joules
kilowatt-hrs.	8.605×10^{2}	kg.-calories
kilowatt-hrs.	8.5985×10^{3}	kg.-meters
kilowatt-hrs.	2.275×10^{1}	pounds of water raised from 62° to 212°F.
	L	
links (engineers)	1.2×10^{1}	inches
links (surveyors)	7.92	inches
liters	1.0×10^{3}	cu. cm.
liters	6.102×10^{1}	cu. inches
liters	1.0×10^{-3}	cu. meters
liters	2.642×10^{-1}	gallons (u.s. liquid)
liters	2.113	pints (u.s. liquid)
liters	1.057	quarts (u.s. liquid)

FIGURE A.136　Measurement conversions *(continued)*.

To convert	Multiply by	To obtain
	M	
meters	1.0×10^2	centimeters
meters	3.281	feet
meters	3.937×10^1	inches
meters	1.0×10^{-3}	kilometers
meters	5.396×10^{-4}	miles (nautical)
meters	6.214×10^{-4}	miles (statute)
meters	1.0×10^3	millimeters
meters/min.	1.667	cms./sec.
meters/min.	3.281	feet/min.
meters/min.	5.468×10^{-2}	feet/sec.
meters/min.	6.0×10^{-2}	kms./hr.
meters/min.	3.238×10^{-2}	knots
meters/min.	3.728×10^{-2}	miles/hr.
meters/sec.	1.968×10^2	feet/min.
meters/sec.	3.281	feet/sec.
meters/sec.	3.6	kilometers/hr.
meters/sec.	6.0×10^{-2}	kilometers/min.
meters/sec.	2.237	miles/hr.
meters/sec.	3.728×10^{-2}	miles/min.
miles (neutical)	6.076×10^3	feet
miles (statute)	5.280×10^3	feet
miles/hr.	8.8×10^1	ft./min.
millimeters	1.0×10^{-1}	centimeters
millimeters	3.281×10^{-3}	feet
millimeters	3.937×10^{-2}	inches
millimeters	1.0×10^{-1}	meters
minutes (time)	9.9206×10^{-5}	weeks
	O	
ounces	2.8349×10^1	grams
ounces	6.25×10^{-2}	pounds
ounces (fluid)	1.805	cu. inches
ounces (fluid)	2.957×10^{-2}	liters
	P	
parts/million	5.84×10^{-2}	grains/u.s. gal.
parts/million	7.016×10^{-2}	grains/imp. gal.
parts/million	8.345	pounds/million gal.
pints (liquid)	4.732×10^2	cubic cms.
pints (liquid)	1.671×10^{-2}	cubic ft.
pints (liquid)	2.887×10^1	cubic inches
pints (liqui)	4.732×10^{-4}	cubic meters
pints (liquid)	1.25×10^{-1}	gallons
pints (liquid)	4.732×10^{-1}	liters
pints (liquid)	5.0×10^{-1}	quarts (liquid)

FIGURE A.136   Measurement conversions *(continued).*

To convert	Multiply by	To obtain
pounds	2.56×10^2	drams
pounds	4.448×10^5	dynes
pounds	7.0×10^1	grains
pounds	4.5359×10^2	grams
pounds	4.536×10^{-1}	kilograms
pounds	1.6×10^1	ounces
pounds	3.217×10^1	pounds
pounds	1.21528	pounds (troy)
pounds of water	1.602×10^{-2}	cu. ft.
pounds of water	2.768×10^1	cu. inches
pounds of water	1.198×10^{-1}	gallons
pounds of water/min.	2.670×10^{-4}	cu. ft./sec.
pound-feet	1.356×10^7	cm.-dynes
pound-feet	1.3825×10^4	cm.-grams
pound-feet	1.383×10^{-1}	meter-kgs.
pounds/cu. ft.	1.602×10^{-2}	grams/cu. cm.
pounds/cu. ft.	5.787×10^{-4}	pounds/cu. inches
pounds/sq. in.	6.804×10^{-2}	atmospheres
pounds/sq. in.	2.307	feet of water
pounds/sq. in.	2.036	inches of mercury
pounds/sq. in.	7.031×10^2	kgs./sq. meter
pounds/sq. in.	1.44×10^2	pounds/sq. ft.

Q

To convert	Multiply by	To obtain
quarts (dry)	6.72×10^1	cu. inches
quarts (liquid)	9.464×10^2	cu. cms.
quarts (liquid)	3.342×10^{-2}	cu. ft.
quarts (liquid)	5.775×10^1	cu. inches
quarts (liquid)	2.5×10^{-1}	gallons

R

To convert	Multiply by	To obtain
revolutions	3.60×10^2	degrees
revolutions	4.0	quadrants
rods (surveyors' meas.)	5.5	yards
rods	1.65×10^1	feet
rods	1.98×10^2	inches
rods	3.125×10^{-3}	miles

FIGURE A.136 Measurement conversions *(continued)*.

To convert	Multiply by	To obtain
	S	
slugs	3.217×10^1	pounds
square centimeters	1.076×10^{-3}	sq. feet
square centimeters	1.550×10^{-1}	sq. inches
square centimeters	1.0×10^{-4}	sq. meters
square centimeters	1.0×10^2	sq. millimeters
square feet	2.296×10^{-5}	acres
square feet	9.29×10^2	sq. cms.
square feet	1.44×10^2	sq. inches
square feet	9.29×10^{-2}	sq. meters
square feet	3.587×10^{-3}	sq. miles
square inches	6.944×10^{-3}	sq. ft.
square inches	6.452×10^2	sq. millimeters
square miles	6.40×10^2	acres
square miles	2.788×10^7	sq. ft.
square yards	2.066×10^{-4}	acres
square yards	8.361×10^3	sq. cms.
square yards	9.0	sq. ft.
square yards	1.296×10^3	sq. inches
	T	
temperature (°C.) +273	1.0	absolute temperature (°K.)
temperature (°C.) +17.78	1.8	temperature (°F.)
temperature (°F.) +460	1.0	absolute temperature (°R.)
temperature (°F.) −32	$\frac{8}{9}$	temperature (°C.)
tons (long)	2.24×10^2	pounds
tons (long)	1.12	tons (short)
tons (metric)	2.205×10^5	pounds
tons (short)	2.0×10^3	pounds
	W	
watts	3.4129	btu/hr.
watts	5.688×10^{-2}	btu/min.
watts	4.427×10^1	ft.-lbs/min.
watts	7.378×10^{-1}	ft.-lbs./sec
watts	1.341×10^{-3}	horsepower
watts	1.36×10^{-3}	horsepower (metric)
watts	1.0×10^{-3}	kilowatts
watt-hours	3.413	btu
watt-hours	2.656×10^3	foot-lbs.
watt-hours	1.341×10^{-3}	horsepower-hours
watt (international)	1.000165	watt (absolute)
weeks	1.68×10^2	hours
weeks	1.008×10^4	minutes
weeks	6.048×10^5	seconds

Source: *Pump Handbook* by I. J. Karassik et al. Copyright 1976, McGraw-Hill, Inc.

FIGURE A.136   Measurement conversions *(continued).*

Contaminant	Suggested maximum level, mg/L
Calcium	2 (0.1 meq/L)
Magnesium	4 (0.3 meq/L)
Sodium	70 (3 meq/L)
Potassium	8 (0.2 meq/L)
Fluoride	0.2
Chlorine	0.5
Chloramines	0.1
Nitrate (N)	2
Sulfate	100
Copper, barium, zinc	0.1 each
Arsenic, lead, silver	0.005 each
Chromium	0.014
Cadmium	0.001
Selenium	0.09
Aluminum	0.01
Mercury	0.0002
Bacteria	200 (cfu/mL)

Source: Association for the Advancement of Medical Instrumentation (AAMI) "Hemodialysis Systems Standard," March 1990. Adopted by American National Standards Institute (ANSI), 1992.

FIGURE A.137   AAMI/ANSI water quality standards.

Nom. pipe size, in	Relative humidity, %														
	20			50			70			80			90		
	THK*	HG†	ST‡	THK	HG	ST	THK	HG	ST	THK	HG	ST	THK	HG	ST
0.50				0.5	2	66	0.5	2	66	0.5	2	66	1.0	2	68
0.75				0.5	2	67	0.5	2	67	0.5	2	67	0.5	2	67
1.00				0.5	3	66	0.5	3	66	0.5	3	66	1.0	2	68
1.25				0.5	3	66	0.5	3	66	0.5	3	66	1.0	3	67
1.50				0.5	4	65	0.5	4	65	0.5	4	65	1.0	3	67
2.00				0.5	5	66	0.5	5	66	0.5	5	66	1.0	3	67
2.50				0.5	5	65	0.5	5	65	0.5	5	65	1.0	4	67
3.00				0.5	7	65	0.5	7	65	0.5	7	65	1.0	4	67
3.50	Condensation			0.5	8	65	0.5	8	65	0.5	8	65	1.0	4	68
4.00	control not			0.5	8	65	0.5	8	65	0.5	8	65	1.0	5	67
5.00	required for this			0.5	10	65	0.5	10	65	0.5	10	65	1.0	6	67
6.00	condition			0.5	12	65	0.5	12	65	0.5	12	65	1.0	7	67
8.00				1.0	9	67	1.0	9	67	1.0	9	67	1.0	9	67
10.00				1.0	11	67	1.0	11	67	1.0	11	67	1.0	11	67
12.00				1.0	12	67	1.0	12	67	1.0	12	67	1.0	12	67

*THK—Insulation thickness, inches.
†HG—Heat gain/lineal foot (pipe) 28 ft (flat).
‡ST—Surface temperature.

FIGURE A.138   Insulation thickness to prevent condensation 500°F and 700°F ambient temperature.

Nom. pipe size, in	Relative humidity, %														
	20			50			70			80			90		
	THK*	HG†	ST‡	THK	HG	ST	THK	HG	ST	THK	HG	ST	THK	HG	ST
0.50				0.5	4	64	0.5	4	64	0.5	4	64	1.5	2	68
0.75				0.5	4	64	0.5	4	64	0.5	4	64	1.5	3	67
1.00				0.5	6	63	0.5	6	63	1.0	4	66	1.5	3	67
1.25				0.5	6	63	0.5	6	63	1.0	5	65	1.5	3	67
1.50				0.5	8	62	0.5	8	62	1.0	5	66	1.5	4	67
2.00				0.5	8	63	0.5	8	63	1.0	6	66	1.5	4	67
2.50				0.5	10	63	0.5	10	63	1.0	6	66	1.5	5	67
3.00				0.5	12	62	0.5	12	62	1.0	8	65	1.5	6	67
3.50	Condensation			0.5	14	61	0.5	14	61	1.0	7	66	1.5	6	67
4.00	control not			0.5	15	62	0.5	15	62	1.0	9	65	1.5	7	67
5.00	required for this			0.5	16	63	0.5	16	63	1.0	11	65	2.0	7	67
6.00	condition			0.5	22	61	0.5	22	61	1.0	13	65	2.0	8	67
8.00				1.0	16	65	1.0	16	65	1.0	16	65	2.0	10	67
10.00				1.0	20	65	1.0	20	65	1.0	20	65	2.0	11	67
12.00				1.0	22	65	1.0	22	65	1.0	22	65	2.0	13	67

*THK—Insulation thickness, inches.
†HG—Heat gain/lineal foot (pipe) 28 ft (flat).
‡ST—Surface temperature.

FIGURE A.139 Insulation thickness to prevent condensation, 340°F service temperature and 700°F ambient temperature.

Type of soil	Required sq. ft. of leaching area/100 gal. (m²/L)	Maximum absorption capacity gals./sq. ft. of leaching area for a 24 hr. period (L/m²)
1. Coarse sand or gravel	20 (.005)	5 (203.7)
2. Fine sand	25 (.006)	4 (162.9)
3. Sandy loam or sandy clay	40 (.010)	2.5 (101.9)
4. Clay with considerable sand or gravel	90 (.022)	1.10 (44.8)
5. Clay with small amount of sand or gravel	120 (.029)	0.83 (33.8)

FIGURE A.140 Design criteria of five typical soils.

Anticipated well yield, gpm	Nominal pump bowl size, in	Optimum well casing size, in	Smallest well casing size, in
Less than 100	4	6 I.D.	5 I.D.
75 to 175	5	8 I.D.	6 I.D.
150 to 400	6	10 I.D.	8 I.D.
350 to 650	8	12 I.D.	10 I.D.
600 to 900	10	14 I.D.	12 I.D.
850 to 1300	12	16 I.D.	14 I.D.
1200 to 1800	14	20 I.D.	16 I.D.
1600 to 3000	16	24 I.D.	20 I.D.

FIGURE A.141 Recommended well diameters.

	100°F	200°F	300°F	400°F	500°F	600°F
Fiberglass	0.26	0.30	0.34			
Polyurethane	0.16	0.16	0.16			
Calcium silicate	0.33	0.37	0.41	0.46	0.57	0.60
Cellular glass	0.39	0.47	0.55	0.64	0.74	0.85

Note: These are representative values per inch thickness for one square foot of area. Exact values should be confirmed by the insulation manufacturer.

FIGURE A.142 Heat loss in btus through common insulation materials.

Fixture	Pressure, psi
Basin faucet	8
Basin faucet, self-closing	12
Sink faucet, ⅜ in (0.95 cm)	10
Sink faucet, ½ in (1.3 cm)	5
Dishwasher	15–25
Bathtub faucet	5
Laundry tub cock, ¼ in (0.64 cm)	5
Shower	12
Water closet ball cock	15
Water closet flush valve	15–20
Urinal flush valve	15
Garden hose, 50 ft (15 m), and sill cock	30
Water closet, blowout type	25
Urinal, blowout type	25
Water closet, low-silhouette tank type	30–40
Water closet, pressure tank	20–30

FIGURE A.143 Minimum acceptable operating pressures for various fixtures.

Use	Minimum temperature, °F
Lavatory:	
Hand washing	105
Shaving	115
Showers and tubs	110
Commercial and institutional laundry	180
Residential dishwashing and laundry	140
Commercial spray-type dishwashing as required by National Sanitation Foundation:	
Single or multiple tank hood or rack type:	
Wash	150
Final rinse	180 to 195
Single-tank conveyor type:	
Wash	160
Final rinse	180 to 195
Single-tank rack or door type:	
Single-temperature wash and rinse	165
Chemical sanitizing glasswasher:	
Wash	140
Rinse	75

FIGURE A.144 Minimum hot water temperature for plumbing fixtures and equipment.

		Length of main line in feet (m)								6" (152.4 mm) diameter					
		25	50	75	100	125	150	175	200	225	250	275	300	400	500
		7.6	15.2	22.9	30.5	38.1	45.7	53.3	61	68.6	76.2	83.8	91.4	121.9	152.4
25	7.6	30	24	34	44	54	64	74	84	94	103	113	123	163	168
50	15.2	30	29	39	48	58	68	78	88	98	108	118	128	166	167
75	22.9	30	33	43	53	63	73	83	92	102	112	122	132	164	165
100	30.5	30	37	47	57	67	77	87	97	107	117	127	136	162	163
125	38.1	32	42	52	62	72	81	91	101	111	121	131	141	160	162
150	45.7	36	46	56	66	76	86	96	106	116	125	135	145	159	161
175	53.3	41	51	61	70	80	90	100	110	120	130	140	150	157	159
200	61	45	55	65	75	85	95	105	114	124	134	144	153	156	158
225	68.6	50	59	69	79	89	99	109	119	129	139	149	151	154	157
250	76.2	54	64	74	84	94	103	113	123	133	143	149	150	153	156
275	83.8	58	68	78	88	98	108	118	128	138	146	147	149	152	155
300	91.4	63	73	83	92	102	112	122	132	142	145	146	147	151	154
350	106.7	72	81	91	101	111	121	131	140	141	143	144	145	149	152
400	121.9	80	90	100	110	120	130	136	138	139	141	142	143	147	150
450	137.2	89	99	109	119	129	132	134	136	138	139	141	142	145	149
500	152.4	98	108	118	126	129	131	133	135	136	138	139	140	144	147

Left axis: Length of lateral in feet (m), 3" or 4" (76.2–101.6 mm) diameter

No holding time less than 30 seconds

		Length of main line in feet (m)								8" (203.2 mm) diameter					
		25	50	75	100	125	150	175	200	225	250	275	300	400	500
		7.6	15.2	22.9	30.5	38.1	45.7	53.3	61	68.6	76.2	83.8	91.4	121.9	152.4
25	7.6	30	40	57	75	92	110	128	145	163	180	198	216	223	224
50	15.2	30	44	62	79	97	114	132	150	167	185	202	218	220	221
75	22.9	31	48	66	84	101	119	136	154	172	189	207	214	217	219
100	30.5	35	53	70	88	106	123	141	158	176	194	209	211	214	216
125	38.1	40	57	75	92	110	128	145	163	180	198	206	207	211	214
150	45.7	44	62	79	97	114	132	150	167	185	201	202	204	209	212
175	53.3	48	66	84	101	119	136	154	172	189	197	199	201	206	210
200	61	53	70	88	106	123	141	158	176	192	194	197	199	204	208
225	68.6	57	75	92	110	128	145	163	180	189	192	194	196	202	206
250	76.2	62	79	97	114	132	150	167	183	186	189	191	193	200	204
275	83.8	66	84	101	119	136	154	172	181	184	187	189	191	198	202
300	91.4	70	88	106	123	141	158	174	178	181	184	187	189	196	200
350	106.7	79	97	114	132	150	166	170	174	177	180	183	185	192	197
400	121.9	88	106	123	141	157	162	166	170	174	176	179	181	189	194
450	137.2	97	114	132	148	154	159	163	167	170	173	176	178	186	191
500	152.4	106	123	140	146	151	156	160	164	167	170	173	175	183	189

Left axis: Length of lateral in feet (m), 3" or 4" (76.2–101.6 mm) diameter

No holding time less than 30 seconds

FIGURE A.145   Low pressure air test for building sewers.

Length of lateral in feet (m) 6" (152.4 mm) diameter		Length of main line in feet (m)								8" (203.2 mm) diameter					
		25 7.6	50 15.2	75 22.9	100 30.5	125 38.1	150 45.7	175 53.3	200 61	225 68.6	250 76.2	275 83.8	300 91.4	400 121.9	500 152.4
25	7.6	30	45	63	80	98	116	133	151	168	186	204	221	224	225
50	15.2	37	55	73	90	108	126	143	161	178	196	214	220	222	223
75	22.9	47	65	83	100	118	135	153	171	188	206	217	217	220	221
100	30.5	57	75	93	110	128	145	163	181	198	214	214	215	218	220
125	38.1	67	85	102	120	138	155	173	190	208	211	212	213	216	218
150	45.7	77	95	112	130	148	165	182	200	207	209	210	211	214	217
175	53.3	87	105	122	140	157	175	192	204	206	207	208	209	214	215
200	61	97	114	132	150	167	185	201	202	204	205	206	207	211	214
225	68.6	107	124	142	160	177	195	199	201	203	204	205	206	210	213
250	76.2	117	134	152	169	187	195	198	199	201	202	203	204	209	212
275	83.8	127	144	162	179	192	194	196	198	200	201	202	204	208	210
300	91.4	136	154	172	187	190	192	195	196	198	200	201	202	207	209
350	106.7	156	174	181	185	187	190	193	194	196	198	199	200	205	208
400	121.9	173	178	181	184	186	189	191	192	194	196	197	198	203	206
450	137.2	173	177	180	183	185	187	189	190	192	194	195	196	201	204
500	152.4	173	177	180	182	184	186	188	189	191	192	193	194	200	203

No holding time less than 30 seconds

Length of lateral in feet (m) 4" (101.6 mm) diameter		Length of main line in feet (m)								6" (152.4 mm) diameter					
		25 7.6	50 15.2	75 22.9	100 30.5	125 38.1	150 45.7	175 53.3	200 61	225 68.6	250 76.2	275 83.8	300 91.4	400 121.9	500 152.4
25	7.6	32	59	87	114	142	169	197	224	252	277	277	278	279	280
50	15.2	36	64	91	119	146	174	201	229	256	271	272	273	275	277
75	22.9	41	68	96	123	151	178	206	233	261	265	267	268	272	274
100	30.5	45	73	100	128	155	183	210	238	258	260	262	264	268	271
125	38.1	50	77	105	132	160	187	214	242	253	255	257	259	264	268
150	45.7	54	81	109	136	164	191	219	244	248	251	253	255	261	265
175	53.3	58	86	113	141	168	196	223	239	243	246	249	251	258	262
200	61	63	90	118	145	173	200	228	235	239	242	245	248	255	260
225	68.6	67	95	122	150	177	205	226	231	235	239	242	244	252	257
250	76.2	72	99	127	154	182	209	222	227	231	235	238	241	249	255
275	83.8	76	103	131	158	186	211	218	223	228	231	235	238	247	253
300	91.4	80	108	135	163	190	208	214	220	224	228	232	235	244	250
350	106.7	89	117	144	172	194	201	208	213	218	222	226	229	239	246
400	121.9	98	125	153	179	188	196	202	208	213	217	221	224	235	242
450	137.2	107	134	162	174	183	191	197	203	208	212	216	220	230	238
500	152.4	116	143	160	170	179	186	193	198	203	208	212	215	226	235

No holding time less than 30 seconds

FIGURE A.145 Low pressure air test for building sewers *(continued)*.

		Length of main line in feet (m)								10" (254 mm) diameter					
		25 7.6	50 15.2	75 22.9	100 30.5	125 38.1	150 45.7	175 53.3	200 61	225 68.6	250 76.2	275 83.8	300 91.4	400 121.9	500 152.4
25	7.6	37	65	92	120	147	175	202	230	257	277	278	278	279	280
50	15.2	47	75	102	130	157	185	212	240	267	271	272	273	276	277
75	22.9	57	85	112	140	167	195	222	250	265	266	267	269	272	274
100	30.5	67	95	122	150	177	205	232	257	260	262	263	265	269	271
125	38.1	77	105	132	160	187	215	242	253	255	257	259	261	266	269
150	45.7	87	114	142	169	197	224	245	248	251	254	256	257	263	266
175	53.3	97	124	152	179	207	234	241	245	248	250	252	254	260	264
200	61	107	134	162	189	217	233	237	241	244	247	249	251	258	262
225	68.6	117	144	172	199	225	230	234	238	241	244	246	248	255	260
250	76.2	127	154	182	209	222	227	231	235	238	241	243	246	253	258
275	83.8	136	164	191	213	219	224	229	232	236	238	241	243	251	256
300	91.4	146	174	201	211	217	222	226	230	233	236	239	241	249	254
350	106.7	166	192	200	207	212	217	222	226	229	232	235	237	245	250
400	121.9	181	190	197	203	209	214	218	222	225	228	231	233	241	247
450	137.2	180	188	195	201	206	211	215	218	222	225	227	230	238	244
500	152.4	179	186	193	198	203	208	212	215	219	222	224	227	235	241

Length of lateral in feet (m) 6" (152.4 mm) diameter

No holding time less than 30 seconds

		Length of main line in feet (m)								10" (254 mm) diameter					
		25 7.6	50 15.2	75 22.9	100 30.5	125 38.1	150 45.7	175 53.3	200 61	225 68.6	250 76.2	275 83.8	300 91.4	400 121.9	500 152.4
25	7.6	45	73	100	128	155	183	210	238	265	279	280	280	281	281
50	15.2	63	90	118	145	173	200	228	255	275	275	276	277	278	279
75	22.9	80	108	135	163	190	218	245	270	272	272	273	274	276	277
100	30.5	98	125	153	180	208	235	263	267	268	269	270	271	274	275
125	38.1	116	143	71	198	226	253	263	265	266	267	268	269	272	274
150	45.7	133	161	188	216	243	258	260	262	264	265	266	267	270	272
175	53.3	151	178	206	233	254	256	258	260	262	263	264	265	268	272
200	61	168	196	223	249	252	254	256	258	260	261	262	263	267	269
225	68.6	186	213	241	247	250	253	255	257	258	259	261	262	265	268
250	76.2	204	231	242	246	249	251	253	255	256	258	259	260	264	267
275	83.8	221	237	241	244	247	250	252	254	255	256	258	259	263	266
300	91.4	232	237	240	243	246	249	251	253	254	255	256	258	262	265
350	106.7	232	235	239	242	244	247	249	251	252	253	254	256	260	263
400	121.9	231	234	238	240	243	245	247	249	250	251	253	254	258	261
450	137.2	230	234	237	239	241	243	245	247	248	250	251	252	256	259
500	152.4	230	233	236	238	240	242	244	246	247	249	250	251	255	258

Length of lateral in feet (m) 8" (203.2 mm) diameter

No holding time less than 30 seconds

FIGURE A.145 Low pressure air test for building sewers *(continued)*.

Length of lateral in feet (m) 4" (101.6 mm) diameter		Length of main line in feet (m)								12" (304.8 mm) diameter					
		25	50	75	100	125	150	175	200	225	250	275	300	400	500
		7.6	15.2	22.9	30.5	38.1	45.7	53.3	61	68.6	76.2	83.8	91.4	121.9	152.4
25	7.6	44	84	123	163	202	242	282	321	332	333	334	334	336	336
50	15.2	48	88	128	167	207	246	286	323	324	326	327	328	331	333
75	22.9	53	92	132	172	211	251	290	316	317	319	321	323	327	329
100	30.5	57	97	136	176	216	255	295	308	311	313	316	317	323	326
125	38.1	62	101	141	180	220	260	297	301	304	308	310	312	319	323
150	45.7	66	106	145	185	224	264	290	295	299	302	305	308	315	319
175	53.3	70	110	150	189	229	268	283	289	293	297	300	303	311	316
200	61	75	114	154	194	233	271	277	283	288	292	296	299	308	313
225	68.6	79	119	158	198	238	265	272	278	283	288	291	295	304	310
250	76.2	84	123	163	202	242	259	267	273	278	283	287	291	301	308
275	83.8	88	128	167	207	244	254	262	269	274	279	283	287	298	305
300	91.4	92	132	172	211	239	249	257	264	270	275	279	283	295	302
350	106.7	101	141	180	218	231	241	249	256	262	268	272	276	289	297
400	121.9	110	150	189	210	223	233	242	249	255	261	266	270	283	292
450	137.2	119	158	189	204	216	227	235	243	249	255	260	264	278	288
500	152.4	128	166	184	198	210	221	229	237	243	249	254	259	273	283

No holding time less than 30 seconds

Length of lateral in feet (m) 6" (154.4 mm) diameter		Length of main line in feet (m)								6" (152.4 mm) diameter					
		25	50	75	100	125	150	175	200	225	250	275	300	400	500
		7.6	15.2	22.9	30.5	38.1	45.7	53.3	61	68.6	76.2	83.8	91.4	121.9	152.4
25	7.6	50	89	129	168	208	248	287	327	331	332	333	333	335	336
50	15.2	59	99	139	178	218	257	297	321	323	325	326	327	330	332
75	22.9	69	109	149	188	228	267	307	314	316	318	320	321	326	328
100	30.5	79	119	158	198	238	277	302	306	309	312	314	316	321	325
125	38.1	89	129	168	208	248	287	295	300	303	306	309	311	317	321
150	45.7	99	139	178	218	257	284	289	294	298	301	304	306	314	318
175	53.3	109	149	188	228	267	278	284	289	293	296	299	302	310	315
200	61	119	158	198	238	265	272	278	284	288	292	295	298	306	312
225	68.6	129	168	208	248	260	268	274	279	284	288	291	294	303	309
250	76.2	139	178	218	246	255	263	269	275	280	284	287	290	300	306
275	83.8	149	188	228	242	251	259	266	271	276	280	284	287	297	304
300	91.4	158	198	227	238	248	255	262	268	272	277	281	284	294	301
350	106.7	178	208	221	232	241	249	255	261	266	271	274	278	289	296
400	121.9	189	204	217	227	236	243	250	256	261	265	269	273	284	292
450	137.2	187	201	213	223	231	239	245	251	256	260	264	268	279	288
500	152.4	186	199	210	219	227	234	240	246	251	256	260	263	275	284

No holding time less than 30 seconds

ADOPTED: 1976
REVISED: 1982
REAFFIRMED: 1984

FIGURE A.145 Low pressure air test for building sewers *(continued)*.

Length of line		3–4 in.	6 in.	8 in.	10 in.	12 in.
(feet)	(m)	76.2–101.6 mm	152.4 mm	203.2 mm	254 mm	304.8 mm
25	7.6	30	30	30	30	30
50	15.2	30	30	35	55	79
75	22.9	30	30	53	83	119
100	30.5	30	40	70	110	158
125	38.1	30	50	88	138	198
150	45.7	30	59	106	165	238
175	53.3	31	69	123	193	277
200	61	35	79	141	220	317
225	68.6	40	89	158	248	340
250	76.2	44	99	176	275	340
275	83.8	48	109	194	283	340
300	91.4	53	119	211	283	340
325	99.1	57	129	227	283	340
350	106.7	62	139	227	283	340
375	114.3	66	148	227	283	340
400	121.9	70	158	227	283	362
450	137.2	79	170	227	283	407
500	152.4	88	170	227	314	452

Minimum time, in seconds, for pressure to drop from 3½ (24 kPa) to 2½ (17 kPa) PSIG

FIGURE A.146   Sample air test table.

PVC-DWV TYPE I THERMAL EXPANSION TABLE
Chart Shows Length Change in Inches vs. Degrees Temperature Change
Coefficient of Linear Expansion: $e = 2.9 \times 10^{-5}$ in/in °F

Length (feet)	40°F	50°F	60°F	70°F	80°F	90°F	100°F
20	.278	.348	.418	.487	.557	.626	.696
40	.557	.696	.835	.974	1.114	1.235	1.392
60	.835	1.044	1.253	1.462	1.670	1.879	2.088
80	1.134	1.392	1.670	1.949	2.227	2.506	2.784
100	1.392	1.740	2.088	2.436	2.784	3.132	3.480

PVC-DWV TYPE I THERMAL EXPANSION TABLE (Metric)
Chart Shows Length Change in Millimeters vs. Degrees Temperature Change
Coefficient of Linear Expansion: $\dfrac{.2 \text{ mm}}{\text{mm °C}}$

Length (m)	4.4°C	10°C	15.6°C	21.1°C	26.7°C	32.2°C	37.8°C
6.1	7.1	8.8	10.6	12.4	14.2	15.9	17.7
12.2	14.2	17.7	21.2	24.7	28.3	31.4	35.4
18.3	21.2	26.5	31.8	37.1	42.4	47.7	53.0
24.4	28.8	35.4	42.4	49.5	56.6	63.7	70.7
30.5	35.4	44.2	53.0	61.9	70.7	79.6	88.4

FIGURE A.147   PVC-DWV type 1 thermal expansion table.

Pipe	Fittings		Maximum working pressure
	Schedule	Sizes	
160 psi (SDR 26) (1102.4 kPa)	40	½″ thru 8″ incl. (12.7 mm–203.2 mm)	160 psi–1102.4 kPa
	80	½″ thru 8″ incl. (12.7 mm–203.2 mm)	160 psi–1102.4 kPa
200 psi (SDR 21) (1378 kPa)	40	½″ thru 4″ incl. (12.7 mm–101.6 mm)	200 psi–1378 kPa
	80	½″ thru 8″ incl. (12.7 mm–203.2 mm)	200 psi–1378 kPa
250 psi (SDR 17) (1722.5 kPa)	40	½″ thru 3″ incl. (12.7 mm–76.2 mm)	250 psi–1722.5 kPa
	80	½″ thru 8″ incl. (12.7 mm–101.6 mm)	250 psi–1722.5 kPa
315 psi (SDR 13.5) (2170.4 kPa)	40	½″ thru 1½″ incl. (12.7 mm–38.1 mm)	315 psi–2170.4 kPa
	80	½″ thru 4″ incl. (12.7 mm–101.6 mm)	315 psi–2170.4 kPa
Schedule 40	40 80	½″ thru 1½″ incl. (12.7 mm–38.1 mm)	320 psi–2204.8 kPa
	40 80	2″ thru 4″ incl. (50.8 mm–101.6 mm)	220 psi–1515.8 kPa
	40	5″ thru 8″ incl.	160 psi–1102.4 kPa
Schedule 80	40	½″ thru 1½″ incl. (12.7 mm–38.1 mm)	320 psi–2204.8 kPa
	40	2″ thru 4″ incl. (50.8 mm–101.6 mm)	220 psi–1515.8 kPa
	40	5″ thru 8″ incl.	160 psi–1102.4 kPa
	80	½″ thru 4″ incl. (12.7 mm–101.6 mm)	320 psi–2204.8 kPa
	80	5″ thru 8″ incl. (127 mm–203.2 mm)	250 psi–1722.5 kPa

FIGURE A.148 Maximum working pressure.

Required sq. ft. of leaching area/100 gals septic tank capacity		Maximum septic tank size allowable	
	(m²/L)		(liters)
20–25	(.005–.006)	7500	(28387.5)
40	(.010)	5000	(18925)
90	(.022)	3500	(13247.5)
120	(.030)	3000	(11355)

FIGURE A.149 Septic system requirements.

Because of the many variables encountered, it is not possible to set absolute values for waste/sewage flow rates for all situations. The designer should evaluate each situation and, if figures in this Table need modification, they should be made with the concurrence of the Administrative Authority.

Type of occupancy	Unit gallons (liters) per day
1. Airports	15 (56.8) per employee
	5 (18.9) per passenger
2. Auto washers	Check with equipment manufacturer
3. Bowling alleys (snack bar only)	75 (283.9) per lane
4. Camps:	
Campground with central	
comfort station	35 (132.5) per person
with flush toilets, no showers	25 (94.6) per person
Day camps (no meals served)	15 (56.8) per person
Summer and seasonal	50 (189.3) per person

FIGURE A.150   Estimated waste/sewage flow rates.

THRUST AT FITTINGS IN POUNDS AT 100 psi

Pipe size inches	90° bends	45° bends	22½° bends	Dead ends & tees
1½	415	225	115	295
2	645	350	180	455
2½	935	510	260	660
3	1,395	755	385	985
3½	1,780	962	495	1,260
4	2,295	1,245	635	1,620
5	3,500	1,900	975	2,490
6	4,950	2,710	1,385	3,550
8	8,300	4,500	2,290	5,860
10	12,800	6,900	3,540	9,050
12	18,100	9,800	5,000	12,800

THRUST AT FITTINGS IN PASCALS AT 689 kPa OF WATER PRESSURE

(mm)	90° bends	45° bends	22½° bends	Dead ends & tees
38.1	1846.8	1001.3	511.8	1312.8
50.8	2870.3	1557.5	801	2024.8
63.5	4160.8	2269.5	1157	3937
76.2	6207.8	3359.8	1713.3	4383.3
88.9	7921	4280.9	2202.8	5607
101.6	10212.8	5540.3	2815.8	7209
127	15575	8455	4338.8	11080.5
152.4	22027.5	12059.5	6163.3	15797.5
203.2	36935	20025	10190.5	26077
254	56960	30705	15753	40272.5
304.8	80545	43610	22250	56960

FIGURE A.151   Thrust at fittings in pounds and pascals.

Single family dwellings— number of bedrooms	Multiple dwelling units or apartments—one bedroom each	Other uses: maximum fixture units served	Minimum septic tank capacity in gallons (liters)
1 or 2		15	750 (2838)
3		20	1000 (3785)
4	2 units	25	1200 (4542)
5 or 6	3	33	1500 (5677.5)
	4	45	2000 (7570)
	5	55	2250 (8516.3)
	6	60	2500 (9462.5)
	7	70	2750 (10408.8)
	8	80	3000 (11355)
	9	90	3250 (12301.3)
	10	100	3500 (13247.5)

Extra bedroom, 150 gallons (567.8 liters) each.
Extra dwelling units over 10, 250 gallons (946.3 liters) each.
Extra fixture units over 100, 25 gallons (94.6 liters) per fixture units.

*Note: Septic tank sizes in this table include sludge storage capacity and the connection of domestic food waste disposal units without further volume increase.

FIGURE A.152 Capacity of septic tanks.

METRIC SYSTEM
(INTERNATIONAL SYSTEM OF UNITS – SI)
(Continued)

TO CONVERT	INTO	MULTIPLY BY
Liters	Gallons (U.S. liquid)	0.2642
Meters	Feet	3.281
Meters	Inches	39.37
Meters	Yards	1.094
Meters/second	Feet/second	3.281
Meters/second	Miles/hr	2.237
Miles (statute)	Kilometers	1.609
Miles/hour	Meters/minute	26.82
Millimeters	Inches	0.03937
Ounces (fluid)	Liters	0.02957
Pints (liquid)	Cubic centimeters	473.2
Pounds	Kilograms	0.4536
PSI	Pascals	6,895
Quarts (liquid)	Liters	0.9463
Radians	Degrees	57.30
Square inches	Square millimeters	645.2
Square meters	Square inches	1,550
Square millimeters	Square inches	1.550×10^{-3}
Watts	Btu/hour	3.4129
Watts	Btu/minute	0.05688
Watts	Foot-pounds/second	0.7378
Watts	Horsepower	1.341×10^{-3}

FIGURE A.153 Metric conversions. *(Reprinted from the 2000 Uniform Plumbing Code (UPC) with the permission of the International Association of Plumbing and Mechanical Officials (IAPMO).*

METRIC SYSTEM
(INTERNATIONAL SYSTEM OF UNITS – SI)

TO CONVERT	INTO	MULTIPLY BY
Atmospheres	Cms of mercury	76.0
Btu	Joules	1,054.8
Btu/hour	Watts	0.2931
Btu/minute	Kilowatts	0.01757
Btu/minute	Watts	17.57
Centigrade	Fahrenheit	(°C x 9/5) + 32°
Circumference	Radians	6.283
Cubic centimeters	Cubic inches	0.06102
Cubic feet	Cubic meters	0.02832
Cubic feet	Liters	28.32
Cubic feet/minute	Cubic cms/second	472.0
Cubic inches	Cubic cms	16.39
Cubic inches	Liters	0.01639
Cubic meters	Gallons (U.S. liquid)	264.2
Feet	Centimeters	30.48
Feet	Meters	0.3048
Feet	Millimeters	304.8
Feet of water	Kgs/square cm	0.03048
Foot-Pounds	Joules	1.356
Foot-pounds/minute	Kilowatts	2.260×10^{-5}
Foot-pounds/second	Kilowatts	1.356×10^{-3}
Gallons	Liters	3.785
Horsepower	Watts	745.7
Horsepower-hours	Joules	2.684×10^{6}
Horsepower-hours	Kilowatt-hours	0.7457
Joules	Btu	9.480×10^{-4}
Joules	Foot-pounds	0.7376
Joules	Watt-hours	2.778×10^{-4}
Kilograms	Pounds	2.205
Kilograms	Tons (short)	1.102×10^{-3}
Kilometers	Miles	0.6214
Kilometers/hour	Miles/hour	0.6214
Kilowatts	Horsepower	1.341
Kilowatt-hours	Btu	3,413
Kilowatt-hours	Foot-pounds	2.655×10^{6}
Kilowatt-hours	Joules	3.6×10^{6}
Liters	Cubic feet	0.03531

FIGURE A.154 Metric conversions. *(Reprinted from the 2000 Uniform Plumbing Code (UPC) with the permission of the International Association of Plumbing and Mechanical Officials (IAPMO).*

Equivalent Length of Pipe for Various Fittings

Diameter of fitting	90° Standard Elbow	45° Standard Elbow	Standard Tee 90°	Coupling or Straight Run of Tee	Gate Valve	Globe Valve	Angle Valve
mm	mm	mm	mm	mm	mm	mm	mm
10	305	183	457	91	61	2438	1219
15	610	366	914	183	122	4572	2438
20	762	457	1219	244	152	6096	3658
25	914	549	1524	274	183	7620	4572
32	1219	732	1829	366	244	10668	5486
40	1524	914	2134	457	305	13716	6706
50	2134	1219	3048	610	396	16764	8534
65	2438	1524	3658	762	488	19812	10363
80	3048	1829	4572	914	610	24384	12192
100	4267	2438	6401	1219	823	38100	16764
125	5182	3048	7620	1524	1006	42672	21336
150	6096	3658	9144	1829	1219	50292	24384

*Allowances are based on non-recessed threaded fittings. Use one-half (1/2) the allowances for recessed threaded fittings or streamline solder fittings.

FIGURE A.155 Metric equivalent pipe lengths. *(Reprinted from the 2000 Uniform Plumbing Code (UPC) with the permission of the International Association of Plumbing and Mechanical Officials (IAPMO).*

Example

Fixture Units and Estimated Demands

	Building Supply Demand				Branch to Hot Water System		
Kind of Fixtures	No. of Fixtures	Fixture Unit Demand	Total Units	Building Supply Demand in gpm (L per sec)	No. of Fixtures	Fixture Unit Demand Calculation	Demand in gallons per minute (L per sec)
Water Closets	130	8.0	1040	–	–	–	–
Urinals	30	4.0	120	–	–	–	–
Shower Heads	12	2.0	24	–	12	12 x 2 x 3/4 = 18	–
Lavatories	100	1.0	100	–	100	100 x 1 x 3/4 = 75	–
Service Sinks	27	3.0	81	–	27	27 x 3 x 3/4 = 61	–
Total			1365	252 gpm (15.8 L/s)		154	55 gpm (3.4 L/s)

Allowing for 15 psi (103.4 kPa) at the highest fixture under the maximum demand of 252 gallons per minute (15.8 L/sec.), the pressure available for friction loss is found by the following:

$$55 - [15 + (45 \times 0.43)] = 20.65 \text{ psi}$$

$$\text{Metric: } 379 - [103.4 + (13.7 \times 9.8)] = 142.3 \text{ kPa}$$

The allowable friction loss per 100 feet (30.4 m) of pipe is therefore:

$$100 \times 20.65 \div 200 = 10.32 \text{ psi}$$

$$\text{Metric: } 30.4 \times 142.3 \div 60.8 = 71.1 \text{ kPa}$$

FIGURE A.156 Estimated demand can be figured with this information. *(Reprinted from the 2000 Uniform Plumbing Code (UPC) with the permission of the International Association of Plumbing and Mechanical Officials (IAPMO).*

-100°–30°		
°C	*Base temperature*	°F
-73	-100	-148
-68	-90	-130
-62	-80	-112
-57	-70	-94
-51	-60	-76
-46	-50	-58
-40	-40	-40
-34.4	-30	-22
-28.9	-20	-4
-23.3	-10	14
-17.8	0	32
-17.2	1	33.8
-16.7	2	35.6
-16.1	3	37.4
-15.6	4	39.2
-15.0	5	41.0
-14.4	6	42.8
-13.9	7	44.6
-13.3	8	46.4
-12.8	9	48.2
-12.2	10	50.0
-11.7	11	51.8
-11.1	12	53.6
-10.6	13	55.4
-10.0	14	57.2

31°–71°		
°C	*Base temperature*	°F
-0.6	31	87.8
0	32	89.6
0.6	33	91.4
1.1	34	93.2
1.7	35	95.0
2.2	36	96.8
2.8	37	98.6
3.3	38	100.4
3.9	39	102.2
4.4	40	104.0
5.0	41	105.8
5.6	42	107.6

FIGURE A.157   Temperature conversion.

Vacuum in inches of mercury	Boiling point
29	76.62
28	99.93
27	114.22
26	124.77
25	133.22
24	140.31
23	146.45
22	151.87
21	156.75
20	161.19
19	165.24
18	169.00
17	172.51
16	175.80
15	178.91
14	181.82
13	184.61
12	187.21
11	189.75
10	192.19
9	194.50
8	196.73
7	198.87
6	200.96
5	202.25
4	204.85
3	206.70
2	208.50
1	210.25

FIGURE A.158   Boiling points of water based on pressure.

LOAD VALUES ASSIGNED TO FIXTURES[a]

FIXTURE	OCCUPANCY	TYPE OF SUPPLY CONTROL	LOAD VALUES, IN WATER SUPPLY FIXTURE UNITS (wsfu)		
			Cold	Hot	Total
Bathroom group	Private	Flush tank	2.7	1.5	3.6
Bathroom group	Private	Flush valve	6.0	3.0	8.0
Bathtub	Private	Faucet	1.0	1.0	1.4
Bathtub	Public	Faucet	3.0	3.0	4.0
Bidet	Private	Faucet	1.5	1.5	2.0
Combination fixture	Private	Faucet	2.25	2.25	3.0
Dishwashing machine	Private	Automatic		1.4	1.4
Drinking fountain	Offices, etc.	$^3/_8"$ valve	0.25		0.25
Kitchen sink	Private	Faucet	1.0	1.0	1.4
Kitchen sink	Hotel, restaurant	Faucet	3.0	3.0	4.0
Laundry trays (1 to 3)	Private	Faucet	1.0	1.0	1.4
Lavatory	Private	Faucet	0.5	0.5	0.7
Lavatory	Public	Faucet	1.5	1.5	2.0
Service sink	Offices, etc.	Faucet	2.25	2.25	3.0
Shower head	Public	Mixing valve	3.0	3.0	4.0
Shower head	Private	Mixing valve	1.0	1.0	1.4
Urinal	Public	1" flush valve	10.0		10.0
Urinal	Public	$^3/_4"$ flush valve	5.0		5.0
Urinal	Public	Flush tank	3.0		3.0
Washing machine (8 lbs.)	Private	Automatic	1.0	1.0	1.4
Washing machine (8 lbs.)	Public	Automatic	2.25	2.25	3.0
Washing machine (15 lbs.)	Public	Automatic	3.0	3.0	4.0
Water closet	Private	Flush valve	6.0		6.0
Water closet	Private	Flush tank	2.2		2.2
Water closet	Public	Flush valve	10.0		10.0
Water closet	Public	Flush tank	5.0		5.0
Water closet	Public or private	Flushometer tank	2.0		2.0

For SI: 1 inch = 25.4 mm, 1 pound = 0.454 kg.

a. For fixtures not listed, loads should be assumed by comparing the fixture to one listed using water in similar quantities and at similar rates. The assigned loads for fixtures with both hot and cold water supplies are given for separate hot and cold water loads and for total load, the separate hot and cold water loads being three-fourths of the total load for the fixture in each case.

FIGURE A.159 Every fixture involved in plumbing has a load value. They are determined here. *(Courtesy of International Code Council, Inc. and International Plumbing Code 2000)*

Table for Estimating Demand

SUPPLY SYSTEMS PREDOMINANTLY FOR FLUSH TANKS			SUPPLY SYSTEMS PREDOMINANTLY FOR FLUSH VALVES		
Load	Demand		Load	Demand	
(Water supply fixture units)	(Gallons per minute)	(Cubic feet per minute)	(Water supply fixture units)	(Gallons per minute)	(Cubic feet per minute)
1	3.0	0.04104			
2	5.0	0.0684			
3	6.5	0.86892			
4	8.0	1.06944			
5	9.4	1.256592	5	15.0	2.0052
6	10.7	1.430376	6	17.4	2.326032
7	11.8	1.577424	7	19.8	2.646364
8	12.8	1.711104	8	22.2	2.967696
9	13.7	1.831416	9	24.6	3.288528
10	14.6	1.951728	10	27.0	3.60936
11	15.4	2.058672	11	27.8	3.716304
12	16.0	2.13888	12	28.6	3.823248
13	16.5	2.20572	13	29.4	3.930192
14	17.0	2.27256	14	30.2	4.037136
15	17.5	2.3394	15	31.0	4.14408
16	18.0	2.90624	16	31.8	4.241024
17	18.4	2.459712	17	32.6	4.357968
18	18.8	2.513184	18	33.4	4.464912
19	19.2	2.566656	19	34.2	4.571856
20	19.6	2.620128	20	35.0	4.6788
25	21.5	2.87412	25	38.0	5.07984
30	23.3	3.114744	30	42.0	5.61356
35	24.9	3.328632	35	44.0	5.88192
40	26.3	3.515784	40	46.0	6.14928
45	27.7	3.702936	45	48.0	6.41664
50	29.1	3.890088	50	50.0	6.684

FIGURE A.160 This table will let a user estimate demand. *(Courtesy of International Code Council, Inc. and International Plumbing Code 2000)*

Table for Estimating Demand—cont'd

SUPPLY SYSTEMS PREDOMINANTLY FOR FLUSH TANKS			SUPPLY SYSTEMS PREDOMINANTLY FOR FLUSH VALVES		
Load	Demand		Load	Demand	
(Water supply fixture units)	(Gallons per minute)	(Cubic feet per minute)	(Water supply fixture units)	(Gallons per minute)	(Cubic feet per minute)
60	32.0	4.27776	60	54.0	7.21872
70	35.0	4.6788	70	58.0	7.75344
80	38.0	5.07984	80	61.2	8.181216
90	41.0	5.48088	90	64.3	8.595624
100	43.5	5.81508	100	67.5	9.0234
120	48.0	6.41664	120	73.0	9.75864
140	52.5	7.0182	140	77.0	10.29336
160	57.0	7.61976	160	81.0	10.82808
180	61.0	8.15448	180	85.5	11.42964
200	65.0	8.6892	200	90.0	12.0312
225	70.0	9.3576	225	95.5	12.76644
250	75.0	10.0260	250	101.0	13.50168
275	80.0	10.6944	275	104.5	13.96956
300	85.0	11.3628	300	108.0	14.43744
400	105.0	14.0364	400	127.0	16.97736
500	124.0	16.57632	500	143.0	19.11624
750	170.0	22.7256	750	177.0	23.66136
1,000	208.0	27.80544	1,000	208.0	27.80544
1,250	239.0	31.94952	1,250	239.0	31.94952
1,500	269.0	35.95992	1,500	269.0	35.95992
1,750	297.0	39.70296	1,750	297.0	39.70296
2,000	325.0	43.446	2,000	325.0	43.446
2,500	380.0	50.7984	2,500	380.0	50.7984
3,000	433.0	57.88344	3,000	433.0	57.88344
4,000	535.0	70.182	4,000	525.0	70.182
5,000	593.0	79.27224	5,000	593.0	79.27224

For SI: 1 gpm = 3.785 L/m, 1 cfm = 0.4719 L/s.

FIGURE A.161 The table for estimating demand for flush tanks and valves. *(Courtesy of International Code Council, Inc. and International Plumbing Code 2000)*

LOSS OF PRESSURE THROUGH TAPS AND TEES IN POUNDS PER SQUARE INCH (psi)

GALLONS PER MINUTE	SIZE OF TAP OR TEE (inches)						
	5/8	3/4	1	1 1/4	1 1/2	2	3
10	1.35	0.64	0.18	0.08			
20	5.38	2.54	0.77	0.31	0.14		
30	12.1	5.72	1.62	0.69	0.33	0.10	
40		10.2	3.07	1.23	0.58	0.18	
50		15.9	4.49	1.92	0.91	0.28	
60			6.46	2.76	1.31	0.40	
70			8.79	3.76	1.78	0.55	0.10
80			11.5	4.90	2.32	0.72	0.13
90			14.5	6.21	2.94	0.91	0.16
100			17.94	7.67	3.63	1.12	0.21
120			25.8	11.0	5.23	1.61	0.30
140			35.2	15.0	7.12	2.20	0.41
150				17.2	8.16	2.52	0.47
160				19.6	9.30	2.92	0.54
180				24.8	11.8	3.62	0.68
200				30.7	14.5	4.48	0.84
225				38.8	18.4	5.6	1.06
250				47.9	22.7	7.00	1.31
275					27.4	7.70	1.59
300					32.6	10.1	1.88

For SI: 1 inch = 25.4 mm, 1 psi = 6.895 kPa, 1 gpm = 3.785 L/m.

FIGURE A.162 Pressure can be lost in taps and tees This examines the numbers. *(Courtesy of International Code Council, Inc. and International Plumbing Code 2000)*

ALLOWANCE IN EQUIVALENT LENGTH OF PIPE FOR FRICTION LOSS IN VALVES AND THREADED FITTINGS (feet)

FITTING OR VALVE	PIPE SIZES (inches)							
	1/2	3/4	1	1 1/4	1 1/2	2	2 1/2	3
45-degree elbow	1.2	1.5	1.8	2.4	3.0	4.0	5.0	6.0
90-degree elbow	2.0	2.5	3.0	4.0	5.0	7.0	8.0	10.0
Tee, run	0.6	0.8	0.9	1.2	1.5	2.0	2.5	3.0
Tee, branch	3.0	4.0	5.0	6.0	7.0	10.0	12.0	15.0
Gate valve	0.4	0.5	0.6	0.8	1.0	1.3	1.6	2.0
Balancing valve	0.8	1.1	1.5	1.9	2.2	3.0	3.7	4.5
Plug-type cock	0.8	1.1	1.5	1.9	2.2	3.0	3.7	4.5
Check valve, swing	5.6	8.4	11.2	14.0	16.8	22.4	28.0	33.6
Globe valve	15.0	20.0	25.0	35.0	45.0	55.0	65.0	80.0
Angle valve	8.0	12.0	15.0	18.0	22.0	28.0	34.0	40.0

For SI: 1 inch = 25.4 mm, 1 foot = 304.8 mm, 1 degree = 0.0175 rad.

FIGURE A.163   This chart examines the allowances involved in friction loss in valves and threaded fittings. *(Courtesy of International Code Council, Inc. and International Plumbing Code 2000)*

PRESSURE LOSS IN FITTINGS AND VALVES EXPRESSED AS EQUIVALENT LENGTH OF TUBE[a] (feet)

NOMINAL OR STANDARD SIZE (inches)	FITTINGS					VALVES			
	Standard Ell		90-Degree Tee						
	90 Degree	45 Degree	Side Branch	Straight Run	Coupling	Ball	Gate	Butterfly	Check
3/8	0.5	—	1.5	—	—	—	—	—	1.5
1/2	1	0.5	2	—	—	—	—	—	2
5/8	1.5	0.5	2	—	—	—	—	—	2.5
3/4	2	0.5	3	—	—	—	—	—	3
1	2.5	1	4.5	—	—	0.5	—	—	4.5
1 1/4	3	1	5.5	0.5	0.5	0.5	—	—	5.5
1 1/2	4	1.5	7	0.5	0.5	0.5	—	—	6.5
2	5.5	2	9	0.5	0.5	0.5	0.5	7.5	9
2 1/2	7	2.5	12	0.5	0.5	—	1	10	11.5
3	9	3.5	15	1	1	—	1.5	15.5	14.5
3 1/2	9	3.5	14	1	1	—	2	—	12.5
4	12.5	5	21	1	1	—	2	16	18.5
5	16	6	27	1.5	1.5	—	3	11.5	23.5
6	19	7	34	2	2	—	3.5	13.5	26.5
8	29	11	50	3	3	—	5	12.5	39

For SI: 1 inch = 25.4 mm, 1 foot = 304.8 mm, 1 degree = 0.0175 rad.

a. Allowances are for streamlined soldered fittings and recessed threaded fittings. For threaded fittings, double the allowances shown in the table. The equivalent lengths presented above are based on a C factor of 150 in the Hazen-Williams friction loss formula. The lengths shown are rounded to the nearest half-foot.

FIGURE A.164   You can determine pressure losses as equivalent lengths from this table. *(Courtesy of International Code Council, Inc. and International Plumbing Code 2000)*

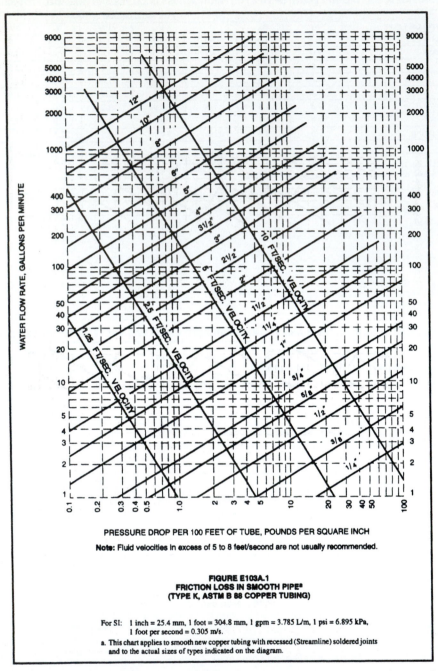

FIGURE E103A.1
FRICTION LOSS IN SMOOTH PIPEª
(TYPE K, ASTM B 88 COPPER TUBING)

For SI: 1 inch = 25.4 mm, 1 foot = 304.8 mm, 1 gpm = 3.785 L/m, 1 psi = 6.895 kPa,
          1 foot per second = 0.305 m/s.
a. This chart applies to smooth new copper tubing with recessed (Streamline) soldered joints
   and to the actual sizes of types indicated on the diagram.

FIGURE A.165   This is one of several tables that determines friction loss. *(Courtesy of
International Code Council, Inc. and International Plumbing Code 2000)*

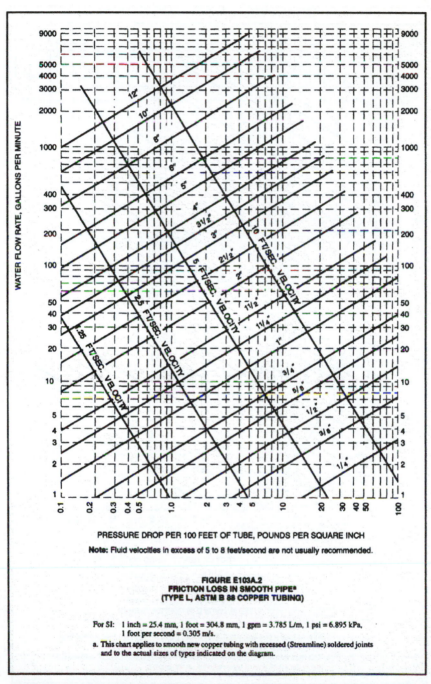

FIGURE A.166 This is one of several tables that determines friction loss. *(Courtesy of International Code Council, Inc. and International Plumbing Code 2000)*

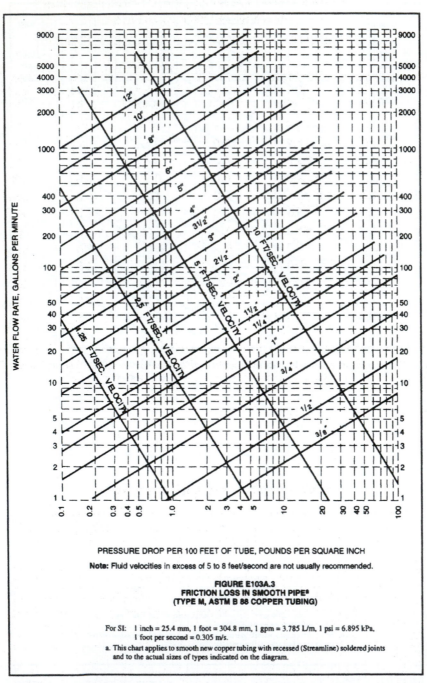

FIGURE A.167 This is one of several tables that determines friction loss. *(Courtesy of International Code Council, Inc. and International Plumbing Code 2000)*

Allowance in equivalent length of pipe for friction loss in valves and threaded fittings.*

Equivalent Length of Pipe for Various Fittings

Diameter of fitting Inches	90° Standard Elbow Feet	45° Standard Elbow Feet	Standard Tee 90° Feet	Coupling or Straight Run of Tee Feet	Gate Valve Feet	Globe Valve Feet	Angle Valve Feet
3/8	1.0	0.6	1.5	0.3	0.2	8	4
1/2	2.0	1.2	3.0	0.6	0.4	15	8
3/4	2.5	1.5	4.0	0.8	0.5	20	12
1	3.0	1.8	5.0	0.9	0.6	25	15
1-1/4	4.0	2.4	6.0	1.2	0.8	35	18
1-1/2	5.0	3.0	7.0	1.5	1.0	45	22
2	7.0	4.0	10.0	2.0	1.3	55	28
2-1/2	8.0	5.0	12.0	2.5	1.6	65	34
3	10.0	6.0	15.0	3.0	2.0	80	40
4	14.0	8.0	21.0	4.0	2.7	125	55
5	17.0	10.0	25.0	5.0	3.3	140	70
6	20.0	12.0	30.0	6.0	4.0	165	80

FIGURE A.168 Pipe length in various fittings in described. *(Reprinted from the 2000 Uniform Plumbing Code (UPC) with the permission of the International Association of Plumbing and Mechanical Officials (IAPMO))*

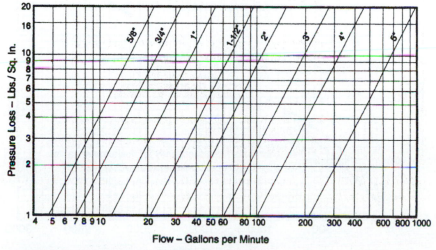

FIGURE A.169 English and metric unit information about friction loss. *(Reprinted from the 2000 Uniform Plumbing Code (UPC) with the permission of the International Association of Plumbing and Mechanical Officials (IAPMO))*

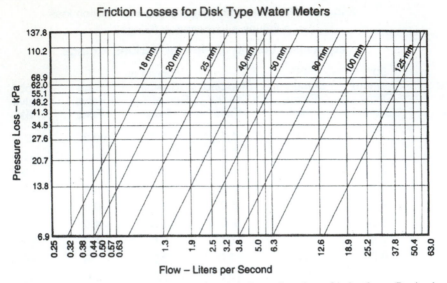

FIGURE A.170 English and metric unit information about friction loss. *(Reprinted from the 2000 Uniform Plumbing Code (UPC) with the permission of the International Association of Plumbing and Mechanical Officials (IAPMO))*

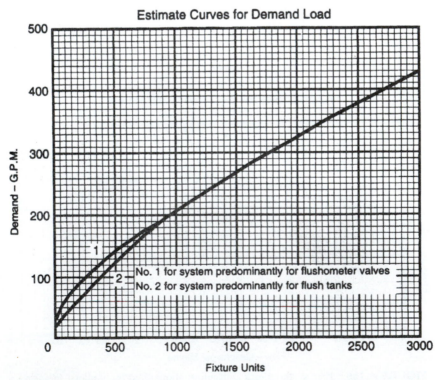

FIGURE A.171 English and metric lengths. *(Reprinted from the 2000 Uniform Plumbing Code (UPC) with the permission of the International Association of Plumbing and Mechanical Officials (IAPMO))*